# REVIEW OF THE ENVIRONMENTAL PROTECTION AGENCY'S DRAFT IRIS ASSESSMENT OF
# FORMALDEHYDE

Committee to Review EPA's Draft IRIS Assessment of Formaldehyde

Board on Environmental Studies and Toxicology

Division on Earth and Life Studies

NATIONAL RESEARCH COUNCIL
*OF THE NATIONAL ACADEMIES*

THE NATIONAL ACADEMIES PRESS
Washington, D.C.
**www.nap.edu**

THE NATIONAL ACADEMIES PRESS    500 Fifth Street, NW    Washington, DC 20001

NOTICE: The project that is the subject of this report was approved by the Governing Board of the National Research Council, whose members are drawn from the councils of the National Academy of Sciences, the National Academy of Engineering, and the Institute of Medicine. The members of the committee responsible for the report were chosen for their special competences and with regard for appropriate balance.

This project was supported by Contract EP-C-09-003 between the National Academy of Sciences and U.S. Environmental Protection Agency. Any opinions, findings, conclusions, or recommendations expressed in this publication are those of the authors and do not necessarily reflect the view of the organizations or agencies that provided support for this project.

International Standard Book Number-13: 978-0-309-21193-2
International Standard Book Number-10: 0-309-21193-X

Additional copies of this report are available from

The National Academies Press
500 Fifth Street, NW
Box 285
Washington, DC 20055

800-624-6242
202-334-3313 (in the Washington metropolitan area)
http://www.nap.edu

Copyright 2011 by the National Academy of Sciences. All rights reserved.

Printed in the United States of America.

# THE NATIONAL ACADEMIES
*Advisers to the Nation on Science, Engineering, and Medicine*

The **National Academy of Sciences** is a private, nonprofit, self-perpetuating society of distinguished scholars engaged in scientific and engineering research, dedicated to the furtherance of science and technology and to their use for the general welfare. Upon the authority of the charter granted to it by the Congress in 1863, the Academy has a mandate that requires it to advise the federal government on scientific and technical matters. Dr. Ralph J. Cicerone is president of the National Academy of Sciences.

The **National Academy of Engineering** was established in 1964, under the charter of the National Academy of Sciences, as a parallel organization of outstanding engineers. It is autonomous in its administration and in the selection of its members, sharing with the National Academy of Sciences the responsibility for advising the federal government. The National Academy of Engineering also sponsors engineering programs aimed at meeting national needs, encourages education and research, and recognizes the superior achievements of engineers. Dr. Charles M. Vest is president of the National Academy of Engineering.

The **Institute of Medicine** was established in 1970 by the National Academy of Sciences to secure the services of eminent members of appropriate professions in the examination of policy matters pertaining to the health of the public. The Institute acts under the responsibility given to the National Academy of Sciences by its congressional charter to be an adviser to the federal government and, upon its own initiative, to identify issues of medical care, research, and education. Dr. Harvey V. Fineberg is president of the Institute of Medicine.

The **National Research Council** was organized by the National Academy of Sciences in 1916 to associate the broad community of science and technology with the Academy's purposes of furthering knowledge and advising the federal government. Functioning in accordance with general policies determined by the Academy, the Council has become the principal operating agency of both the National Academy of Sciences and the National Academy of Engineering in providing services to the government, the public, and the scientific and engineering communities. The Council is administered jointly by both Academies and the Institute of Medicine. Dr. Ralph J. Cicerone and Dr. Charles M. Vest are chair and vice chair, respectively, of the National Research Council.

www.national-academies.org

### COMMITTEE TO REVIEW EPA'S DRAFT IRIS ASSESSMENT OF FORMALDEHYDE

*Members*

**JONATHAN M. SAMET** (*Chair*), University of Southern California, Los Angeles
**ANDREW F. OLSHAN** (*Vice-Chair*), University of North Carolina at Chapel Hill
**JOHN BAILER**, Miami University, Oxford, OH
**SANDRA J.S. BAIRD**, Massachusetts Department of Environmental Protection, Boston
**HARVEY CHECKOWAY**, University of Washington School of Public Health and Community Medicine, Seattle
**RICHARD A. CORLEY**, Pacific Northwest National Laboratory, Richland, WA
**DAVID C. DORMAN**, North Carolina State University, Raleigh
**CHARLES H. HOBBS**, Lovelace Respiratory Research Institute, Albuquerque, NM
**MICHAEL D. LAIOSA**, University of Wisconsin, Milwaukee
**IVAN RUSYN**, University of North Carolina at Chapel Hill
**MARY ALICE SMITH**, University of Georgia, Athens
**LESLIE T. STAYNER**, University of Illinois, Chicago
**HELEN H. SUH**, National Opinion Research Center, University of Chicago, IL
**YILIANG ZHU**, University of South Florida, Tampa
**PATRICK A. ZWEIDLER-MCKAY**, The University of Texas M. D. Anderson Cancer Center, Houston

*Staff*

**ELLEN K. MANTUS**, Project Director
**HEIDI MURRAY-SMITH**, Program Officer
**KERI SCHAFFER**, Research Associate
**NORMAN GROSSBLATT**, Senior Editor
**MIRSADA KARALIC-LONCAREVIC**, Manager, Technical Information Center
**RADIAH ROSE**, Manager, Editorial Projects
**PANOLA GOLSON**, Program Associate

*Sponsor*

**U.S. ENVIRONMENTAL PROTECTION AGENCY**

# BOARD ON ENVIRONMENTAL STUDIES AND TOXICOLOGY[1]

*Members*

ROGENE F. HENDERSON (*Chair*), Lovelace Respiratory Research Institute, Albuquerque, NM
PRAVEEN AMAR, Northeast States for Coordinated Air Use Management, Boston, MA
TINA BAHADORI, American Chemistry Council, Washington, DC
MICHAEL J. BRADLEY, M.J. Bradley & Associates, Concord, MA
DALLAS BURTRAW, Resources for the Future, Washington, DC
JAMES S. BUS, Dow Chemical Company, Midland, MI
JONATHAN Z. CANNON, University of Virginia, Charlottesville
GAIL CHARNLEY, HealthRisk Strategies, Washington, DC
FRANK W. DAVIS, University of California, Santa Barbara
RICHARD A. DENISON, Environmental Defense Fund, Washington, DC
H. CHRISTOPHER FREY, North Carolina State University, Raleigh
J. PAUL GILMAN, Covanta Energy Corporation, Fairfield, NJ
RICHARD M. GOLD, Holland & Knight, LLP, Washington, DC
LYNN R. GOLDMAN, George Washington University, Washington, DC
LINDA E. GREER, Natural Resources Defense Council, Washington, DC
WILLIAM E. HALPERIN, University of Medicine and Dentistry of New Jersey, Newark
PHILIP K. HOPKE, Clarkson University, Potsdam, NY
HOWARD HU, University of Michigan, Ann Arbor
ROGER E. KASPERSON, Clark University, Worcester, MA
THOMAS E. MCKONE, University of California, Berkeley
TERRY L. MEDLEY, E.I. du Pont de Nemours & Company, Wilmington, DE
JANA MILFORD, University of Colorado at Boulder, Boulder
FRANK O'DONNELL, Clean Air Watch, Washington, DC
RICHARD L. POIROT, Vermont Department of Environmental Conservation, Waterbury
DANNY D. REIBLE, University of Texas, Austin
ROBERT F. SAWYER, University of California, Berkeley
KATHRYN G. SESSIONS, Health and Environmental Funders Network, Bethesda, MD
JOYCE S. TSUJI, Exponent Environmental Group, Bellevue, WA
MARK J. UTELL, University of Rochester Medical Center, Rochester, NY

*Senior Staff*

JAMES J. REISA, Director
DAVID J. POLICANSKY, Scholar
RAYMOND A. WASSEL, Senior Program Officer for Environmental Studies
SUSAN N.J. MARTEL, Senior Program Officer for Toxicology
ELLEN K. MANTUS, Senior Program Officer for Risk Analysis
EILEEN N. ABT, Senior Program Officer
RUTH E. CROSSGROVE, Senior Editor
MIRSADA KARALIC-LONCAREVIC, Manager, Technical Information Center
RADIAH ROSE, Manager, Editorial Projects

---

[1] This study was planned, overseen, and supported by the Board on Environmental Studies and Toxicology.

## OTHER REPORTS OF THE
## BOARD ON ENVIRONMENTAL STUDIES AND TOXICOLOGY

Review of the Environmental Protection Agency's Draft IRIS Assessment of Formaldehyde (2011)
The Use of Title 42 Authority at the U.S. Environmental Protection Agency (2010)
Review of the Environmental Protection Agency's Draft IRIS Assessment of Tetrachloroethylene (2010)
Hidden Costs of Energy: Unpriced Consequences of Energy Production and Use (2009)
Contaminated Water Supplies at Camp Lejeune—Assessing Potential Health Effects (2009)
Review of the Federal Strategy for Nanotechnology-Related Environmental, Health, and Safety Research (2009)
Science and Decisions: Advancing Risk Assessment (2009)
Phthalates and Cumulative Risk Assessment: The Tasks Ahead (2008)
Estimating Mortality Risk Reduction and Economic Benefits from Controlling Ozone Air Pollution (2008)
Respiratory Diseases Research at NIOSH (2008)
Evaluating Research Efficiency in the U.S. Environmental Protection Agency (2008)
Hydrology, Ecology, and Fishes of the Klamath River Basin (2008)
Applications of Toxicogenomic Technologies to Predictive Toxicology and Risk Assessment (2007)
Models in Environmental Regulatory Decision Making (2007)
Toxicity Testing in the Twenty-first Century: A Vision and a Strategy (2007)
Sediment Dredging at Superfund Megasites: Assessing the Effectiveness (2007)
Environmental Impacts of Wind-Energy Projects (2007)
Scientific Review of the Proposed Risk Assessment Bulletin from the Office of Management and Budget (2007)
Assessing the Human Health Risks of Trichloroethylene: Key Scientific Issues (2006)
New Source Review for Stationary Sources of Air Pollution (2006)
Human Biomonitoring for Environmental Chemicals (2006)
Health Risks from Dioxin and Related Compounds: Evaluation of the EPA Reassessment (2006)
Fluoride in Drinking Water: A Scientific Review of EPA's Standards (2006)
State and Federal Standards for Mobile-Source Emissions (2006)
Superfund and Mining Megasites—Lessons from the Coeur d'Alene River Basin (2005)
Health Implications of Perchlorate Ingestion (2005)
Air Quality Management in the United States (2004)
Endangered and Threatened Species of the Platte River (2004)

Atlantic Salmon in Maine (2004)
Endangered and Threatened Fishes in the Klamath River Basin (2004)
Cumulative Environmental Effects of Alaska North Slope Oil and Gas
 Development (2003)
Estimating the Public Health Benefits of Proposed Air Pollution
 Regulations (2002)
Biosolids Applied to Land: Advancing Standards and Practices (2002)
The Airliner Cabin Environment and Health of Passengers and Crew (2002)
Arsenic in Drinking Water: 2001 Update (2001)
Evaluating Vehicle Emissions Inspection and Maintenance Programs (2001)
Compensating for Wetland Losses Under the Clean Water Act (2001)
A Risk-Management Strategy for PCB-Contaminated Sediments (2001)
Acute Exposure Guideline Levels for Selected Airborne Chemicals (nine
 volumes, 2000-2010)
Toxicological Effects of Methylmercury (2000)
Strengthening Science at the U.S. Environmental Protection Agency (2000)
Scientific Frontiers in Developmental Toxicology and Risk Assessment (2000)
Ecological Indicators for the Nation (2000)
Waste Incineration and Public Health (2000)
Hormonally Active Agents in the Environment (1999)
Research Priorities for Airborne Particulate Matter (four volumes, 1998-2004)
The National Research Council's Committee on Toxicology: The First 50
 Years (1997)
Carcinogens and Anticarcinogens in the Human Diet (1996)
Upstream: Salmon and Society in the Pacific Northwest (1996)
Science and the Endangered Species Act (1995)
Wetlands: Characteristics and Boundaries (1995)
Biologic Markers (five volumes, 1989-1995)
Science and Judgment in Risk Assessment (1994)
Pesticides in the Diets of Infants and Children (1993)
Dolphins and the Tuna Industry (1992)
Science and the National Parks (1992)
Human Exposure Assessment for Airborne Pollutants (1991)
Rethinking the Ozone Problem in Urban and Regional Air Pollution (1991)
Decline of the Sea Turtles (1990)

*Copies of these reports may be ordered from the National Academies Press*
*(800) 624-6242 or (202) 334-3313*
*www.nap.edu*

# Preface

The U.S. Environmental Protection Agency (EPA) released noncancer and cancer assessments of formaldehyde for its Integrated Risk Information System (IRIS) in 1990 and 1991, respectively. The agency began reassessing formaldehyde in 1998 and released a draft IRIS assessment in June 2010. Much research has been conducted since the original assessments, and scientists are currently debating the carcinogenic properties of formaldehyde and the ways that it might cause cancer. Given the complexity of the issues and the knowledge that the assessment will be used as the basis of regulatory decisions, EPA asked the National Research Council (NRC) to conduct an independent scientific review of the draft IRIS assessment.

In this report, the Committee to Review EPA's Draft IRIS Assessment of Formaldehyde first addresses some general issues associated with the draft IRIS assessment. The committee next focuses on questions concerning specific aspects of the draft assessment, including derivation of the reference concentrations and the cancer unit risk estimates for formaldehyde. The committee closes with recommendations for improving the IRIS assessment of formaldehyde and provides some general comments on the IRIS development process.

The present report has been reviewed in draft form by persons chosen for their diverse perspectives and technical expertise in accordance with procedures approved by the NRC Report Review Committee. The purpose of the independent review is to provide candid and critical comments that will assist the institution in making its published report as sound as possible and to ensure that the report meets institutional standards of objectivity, evidence, and responsiveness to the study charge. The review comments and draft manuscript remain confidential to protect the integrity of the deliberative process. We thank the following for their review of this report: Margit L. Bleecker, Center for Occupational and Environmental Neurology; Claude Emond, Université de Montréal; George L. Delclos, The University of Texas Health Science Center at Houston School of Public Health; Lynn R. Goldman, George Washington University; Ulrike Luderer, University of California, Irvine; Roger O. McClellan, Toxicology and Human Health Risk Analysis; Martha S. Sandy, California Environmental Protection Agency; Jeffrey D. Schroeter, The Hamner Institutes for Health Sciences; Susan J. Simmons, University of North Carolina, Wilmington; Joyce S. Tsuji, Exponent; Elizabeth W. Triche, Brown University; Clifford P. Weisel, University of Medicine and Dentistry of New Jersey; Joseph L. Wiemels, University of California, San Francisco.

Although the reviewers listed above have provided many constructive comments and suggestions, they were not asked to endorse the conclusions or recommendations, nor did they see the final draft of the report before its release. The review of the report was overseen by the review coordinator, Kenneth S. Ramos, University of Louisville Health Science Center, and the review monitor, Frank E. Speizer, Harvard Medical School and Harvard School of Public Health. Appointed by NRC, they were responsible for making certain that an independent examination of the report was carried out in accordance with institutional procedures and that all review comments were carefully considered. Responsibility for the final content of the report rests entirely with the committee and the institution.

The committee gratefully acknowledges Danielle DeVoney, Sue Makris, Peter Preuss, and Kathleen Raffaele, of the U.S. Environmental Protection Agency, and Bruce Fowler, of the Agency for Toxic Substances and Disease Registry, for making presentations to the committee.

The committee is also grateful for the assistance of NRC staff in preparing this report. Staff members who contributed to the effort are Ellen Mantus, project director; Heidi Murray-Smith, program officer; Keri Schaffer, research associate; James Reisa, director of the Board on Environmental Studies and Toxicology; Norman Grossblatt, senior editor; Mirsada Karalic-Loncarevic, manager, Technical Information Center; Radiah Rose, manager, editorial projects; and Panola Golson, program associate.

We thank especially the members of the committee for their efforts throughout the development of this report.

Jonathan M. Samet, *Chair*
Andrew F. Olshan, *Vice-Chair*
Committee to Review EPA's Draft
IRIS Assessment of Formaldehyde

# Contents

**SUMMARY** .................................................................................................... 3

**1  INTRODUCTION** ................................................................................. 16
   Formaldehyde and the Draft Assessment, 16
   The Committee's Task and Approach, 20
   Organization of Report, 22
   References, 23

**2  REVIEW OF METHODS** ..................................................................... 24
   Review of the Methodology of the Draft IRIS Assessment
      of Formaldehyde, 25
   Summary, 28
   References, 28

**3  TOXICOKINETICS AND MODES OF ACTION
    OF FORMALDEHYDE** ........................................................................ 29
   Toxicokinetics, 30
   Carcinogenesis: Has a Mode of Action of Formaldehyde
      Been Identified?, 44
   Use of a Biologically Based Dose-Response Model, 46
   Conclusions and Recommendations, 58
   References, 60

**4  PORTAL-OF-ENTRY HEALTH EFFECTS** ...................................... 64
   Irritation, 65
   Decreased Pulmonary Function, 71
   Noncancer Respiratory Tract Pathology, 74
   Asthma, 78
   Respiratory Tract Cancers, 83
   References, 88

xii                                                                                                     Contents

| | | |
|---|---|---|
| **5** | **SYSTEMIC HEALTH EFFECTS** | **92** |

Immunotoxicity, 93
Neurotoxicity, 97
Reproductive and Developmental Toxicity, 102
Lymphohematopoietic Cancers, 108
References, 114

| | | |
|---|---|---|
| **6** | **REFERENCE CONCENTRATIONS FOR NONCANCER EFFECTS AND UNIT RISKS FOR CANCER** | **118** |

Formaldehyde Reference Concentrations, 119
Formaldehyde Unit Risks For Cancer, 133
Conclusions and Recommendations, 144
References, 146

| | | |
|---|---|---|
| **7** | **A ROADMAP FOR REVISION** | **151** |

Critical Revisions of the Current Draft IRIS Assessment
  of Formaldehyde, 151
Future Assessments and the IRIS Process, 152
References, 166

## APPENDIXES

| | | |
|---|---|---|
| **A** | **BIOGRAPHIC INFORMATION ON THE COMMITTEE TO REVIEW EPA'S DRAFT IRIS ASSESSMENT OF FORMALDEHYDE** | **168** |
| **B** | **WEIGHT-OF-EVIDENCE DESCRIPTIONS FROM U.S. ENVIRONMENTAL PROTECTION AGENCY GUIDELINES** | **174** |

## BOXES, FIGURES, AND TABLES

### BOXES

1-1   Statement of Task, 20

### FIGURES

S-1   Illustration of potential process for identifying an RfC, 13
1-1   Formaldehyde chemical structure, 17
1-2   Formaldehyde concentration in various environments, 18
1-3   Timeline of the development of the draft IRIS assessment, 19
2-1   Elements of the IRIS process, 25
3-1   Schematic representation of the mammalian nasal epithelium, 32

| | |
|---|---|
| 4-1 | Odds ratios for physician-diagnosed asthma in children associated with in-home formaldehyde concentrations in air, 82 |
| 5-1 | Origins of lymphohematopoietic cancers, 109 |
| 5-2 | Relative incidence and estimated annual new diagnoses of common lymphohematopoietic cancer subtypes in the United States, 110 |
| 6-1 | Illustration of EPA's process for deriving a reference concentration for formaldehyde, 120 |
| 6-2 | Illustration of a potential process for identifying an RfC from a full database, 132 |
| 7-1 | New IRIS assessment process, 154 |
| 7-2 | Elements of the key steps in the development of a draft IRIS assessment, 155 |
| 7-3 | Example of an article-selection process, 159 |

**TABLES**

| | |
|---|---|
| 3-1 | Analysis of 3D CFD Models by Kimbell et al. (2001a,b) and Overton et al. (2001) for Rat, Monkey, and Human Airways, 41 |
| 3-2 | Overview of the Conolly et al. BBDR Models, 48 |
| 3-3 | Effects of Different Parameters on Predicted Results of the Conolly et al. BBDR Models, 52 |
| 6-1 | Derivation of Candidate RfCs by EPA, 122 |
| 6-2 | Cancer Unit Risk Estimates for Formaldehyde, 142 |
| 7-1 | Criteria for Determining Causality, 157 |
| 7-2 | Hierarchy for Classifying Strength of Causal Inferences on the Basis of Available Evidence, 157 |

# REVIEW OF THE ENVIRONMENTAL PROTECTION AGENCY'S DRAFT IRIS ASSESSMENT OF
# FORMALDEHYDE

# Summary

Formaldehyde is ubiquitous in indoor and outdoor air, and everyone is exposed to formaldehyde at some concentration daily. Formaldehyde is used to produce a wide array of products, particularly building materials; it is emitted from many sources, including power plants, cars, gas and wood stoves, and cigarettes; it is a natural product in some foods; and it is naturally present in the human body as a metabolic intermediate. Much research has been conducted on the health effects of exposure to formaldehyde, including effects on the upper airway, where formaldehyde is deposited when inhaled, and effects on tissues distant from the site of initial contact.

For more than a decade, the U.S. Environmental Protection Agency (EPA) has been in the process of re-evaluating the health effects of formaldehyde; in June 2010, it released its draft health assessment of formaldehyde for EPA's Integrated Risk Information System (IRIS). Given the complex nature of the assessment and recognition that the assessment will be used as a basis of risk calculations and regulatory decisions, EPA asked the National Research Council (NRC) to conduct an independent scientific review of the draft IRIS assessment and to answer questions related specifically to its derivation of reference concentrations (RfCs) for noncancer effects and unit risk estimates for cancer. In response to EPA's request, NRC convened the Committee to Review EPA's Draft IRIS Assessment of Formaldehyde, which prepared this report.

In addressing its charge,[1] the committee reviewed the draft IRIS assessment provided. It did not perform its own assessment, which would have been beyond its charge. Accordingly, the committee did not conduct its own literature searches, review all relevant evidence, systematically formulate its own conclusions regarding causality, or recommend values for the RfC and unit risk. The committee reviewed the draft IRIS assessment and key literature and determined whether EPA's conclusions were supported on the basis of that assessment and literature.

---

[1] See Chapter 1 for the committee's verbatim statement of task.

## THE DRAFT IRIS ASSESSMENT

Overall, the committee noted some recurring methodologic problems in the draft IRIS assessment of formaldehyde. Many of the problems are similar to those which have been reported over the last decade by other NRC committees tasked with reviewing EPA's IRIS assessments for other chemicals. Problems with clarity and transparency of the methods appear to be a repeating theme over the years, even though the documents appear to have grown considerably in length. In the roughly 1,000-page draft reviewed by the present committee, little beyond a brief introductory chapter could be found on the methods for conducting the assessment. Numerous EPA guidelines are cited, but their role in the preparation of the assessment is not clear. In general, the committee found that the draft was not prepared in a consistent fashion; it lacks clear links to an underlying conceptual framework; and it does not contain sufficient documentation on methods and criteria for identifying evidence from epidemiologic and experimental studies, for critically evaluating individual studies, for assessing the weight of evidence, and for selecting studies for derivation of the RfCs and unit risk estimates. This summary highlights the committee's substantive comments and recommendations that should be considered in revision of the draft IRIS assessment; more detailed comments and recommendations can be found at the conclusions of individual chapters or following the discussions on individual health outcomes.

## Toxicokinetics

The committee reviewed the extensive discussion on toxicokinetics of formaldehyde in the draft IRIS assessment and focused on several key issues: the implications of endogenous formaldehyde, the fate of inhaled formaldehyde, the systemic availability of formaldehyde, the ability of formaldehyde to cause systemic genotoxic effects, and the usefulness of various models.

*Endogenous formaldehyde.* Humans and other animals produce formaldehyde through various biologic pathways as part of normal metabolism. Thus, formaldehyde is normally present at low concentrations in all tissues, cells, and bodily fluids. Although there is some debate regarding interpretation of the analytic measurements, formaldehyde has been measured in exhaled breath and is most likely present normally at a concentration of a few parts per billion. The endogenous production of formaldehyde complicates the assessment of the risk associated with formaldehyde inhalation and remains an important uncertainty in assessing the additional dose received by inhalation, particularly at sites beyond the respiratory tract.

*Fate of inhaled formaldehyde.* Formaldehyde is a highly water-soluble, reactive chemical that has a short biologic half-life. Despite species differences in uptake due to differences in breathing patterns and nasal structures, formaldehyde is absorbed primarily at the site of first contact where it undergoes extensive local metabolism and reactions with macromolecules. Thus, the net result is that inhaled formaldehyde remains predominantly in the respiratory epithelium that lines the airways.

*Systemic availability of formaldehyde.* The issue of whether inhaled formaldehyde can reach the systemic circulation is important in assessing the risk of adverse effects at nonrespiratory sites. The draft IRIS assessment provides divergent statements regarding systemic delivery of formaldehyde that need to be resolved. Specifically, some parts of the draft assume that the high reactivity and extensive nasal absorption of formaldehyde restrict systemic delivery of inhaled formaldehyde so that formaldehyde does not go beyond the upper respiratory tract, and other parts of the draft assume that systemic delivery accounts in part for the systemic effects attributed to formaldehyde exposure.

The committee concludes that the weight of evidence suggests that formaldehyde is unlikely to appear in the blood as an intact molecule except perhaps at concentrations high enough to transiently overwhelm the metabolic capability of the tissue at the site of exposure. Thus, direct evidence of systemic delivery of formaldehyde is generally lacking. Furthermore, it is unlikely that formaldehyde reaches distal sites via its hydrated form, methanediol. Although equilibrium dynamics indicate that methanediol would constitute more than 99.9% of the total free and hydrated formaldehyde, experimental data provide compelling evidence that hydration of formaldehyde does not enhance delivery beyond the portal of entry to distal tissues. Pharmacokinetic modeling also supports that conclusion.

*Systemic genotoxic effects of formaldehyde exposure.* The draft IRIS assessment correctly concludes that formaldehyde is a genotoxic (DNA-reactive) chemical that causes cytogenetic effects, such as mutations. Furthermore, the overall body of evidence suggests that inhaled formaldehyde has a cytogenetic effect that can be detected in peripheral (circulating) blood lymphocytes. However, the committee concludes that data are insufficient to conclude definitively that formaldehyde is causing cytogenetic effects at distant sites. First, the observed effects have occurred in highly exposed people, and extrapolating to more typical environmental exposures is difficult given the uncertainty surrounding the form of the dose-response curve for cytogenetic changes. Second, a mechanism that would explain the occurrence of cytogenetic effects in circulating blood cells has not been established. That gap in mechanistic understanding is particularly problematic because the data strongly suggest that formaldehyde is not available systemically in any reactive form. Thus, the committee can only hypothesize that the observed effects result from an unproven mechanism in portal-of-entry tissues.

*Usefulness of various models.* Computational fluid dynamics (CFD) models have been developed to help to predict the dose to nasal tissues from inhaled formaldehyde. EPA fairly evaluated the models and sources of uncertainty but did not use the models to extrapolate to low concentrations. The committee concludes that the models would be useful for that purpose and recommends that EPA use the CFD models to extrapolate to low concentrations, include the results in the revised IRIS assessment, and explain clearly its use of CFD modeling approaches.

A biologically based dose-response (BBDR) model that has been developed for formaldehyde could be used in the derivation of the unit risk estimates. EPA explored the uncertainties associated with the model and sensitivities of various model components to changes in key parameters and assumptions and, on the basis of those extrapolations, decided not to use the BBDR model in its assessment. Although the committee agrees that EPA's evaluation of the model yielded some important findings on model sensitivity, some of the manipulations are extreme, may not be scientifically justified, and should not have been used as the basis of rejection of the use of the BBDR model in its assessment. Model manipulations that yield results that are implausible or inconsistent with available data should be rejected as a basis for judging the utility of the model.

The primary purposes of a BBDR model are to predict as accurately as possible a response to a given exposure, to provide a rational framework for extrapolations outside the range of experimental data (that is, across doses, species, and exposure routes), and to assess the effect of variability and uncertainty on model parameters. In developing a BBDR model, a model structure and parameter values should be chosen to constrain model predictions within biologic and physical limits, all relevant data should be reconciled with the model, and model predictions should be reconciled with credible outcomes. Thus, it provides a valuable method for predicting the range of plausible responses in a given exposure scenario. Given that the BBDR model for formaldehyde is one of the best-developed BBDR models to date, the positive attributes of BBDR models generally, and the limitations of the human data, the committee recommends that EPA use the BBDR model for formaldehyde in its cancer assessment, compare the results with those described in the draft assessment, and discuss the strengths and weaknesses of each approach.

## Mode of Action for Formaldehyde Carcinogenesis

*Mode of action* is defined as a sequence of key events that describe the biologic pathway from exposure to adverse outcome. Understanding the mode of action is important because it can provide support for conclusions regarding causality, and it can affect how unit risk estimates are calculated. Potential modes of action for formaldehyde carcinogenesis have been debated. EPA based its approach to its cancer assessment primarily on the conclusion that formaldehyde is a genotoxic chemical that causes mutations (a mutagenic mode of action). However, for nasal tumors attributed to formaldehyde exposure, animal data also support a mode of action characterized by regenerative cellular proliferation that results from cytotoxicity. Because multiple modes of action may be operational, the committee recommends that EPA provide additional calculations that factor in regenerative cellular proliferation as a mode of action, compare the results with those presented in the draft assessment, and assess the strengths and weaknesses of each approach.

Little is known about a potential mode of action for hematopoietic cancers, such as leukemias, that have been attributed to formaldehyde exposure and that are assumed to arise from sites distant from the portal of entry. The draft IRIS assessment speculates that formaldehyde could reach the bone marrow and cause the mutagenic effects that lead to the cancers noted. However, despite the use of sensitive and selective analytic methods that are capable of differentiating endogenous exposures from exogenous ones, numerous studies have demonstrated that systemic delivery of formaldehyde is unlikely at concentrations that do not overwhelm metabolism. The draft assessment further speculates that circulating hematopoietic stem cells that percolate the nasal capillary bed or nasal-associated lymphoid tissues may be the target cells for the mutagenic effects that eventually lead to the cancers noted. However, experimental evidence supporting that mechanism is lacking.

### Portal-of-Entry Health Effects

EPA evaluated a wide array of outcomes that the committee chose to characterize as portal-of-entry health effects or systemic health effects.[2] The portal-of-entry effects include irritation, decreased pulmonary function, respiratory tract pathology, asthma, and respiratory tract cancers. Overall, the committee found that the noted outcomes were appropriate to evaluate. EPA identified relevant studies for its assessment, and on the basis of the committee's familiarity with the scientific literature, it does not appear to have overlooked any important study. For a few outcomes, however, as noted below, EPA did not discuss or evaluate literature on mode of action that could have supported its conclusions. Although EPA adequately described the studies, critical evaluations of the strengths and weaknesses of the studies were generally deficient, and clear rationales for many conclusions were not provided. In several cases, the committee would not have advanced a particular study or would have advanced other studies to calculate the candidate RfCs. Comments on the specific outcomes are provided below.

*Irritation.* Formaldehyde has been consistently shown to be an eye, nose, and throat irritant, and EPA used several studies of residential exposure to calculate candidate RfCs. However, the favorable attributes of one particular selected study (Richie and Lehnen 1987)[3] were outweighed by the potential for participant-selection bias, and EPA should not have used it to calculate an RfC. Fur-

---

[2]*Portal-of-entry* effects are defined here as effects that arise from direct interaction of inhaled formaldehyde with the airways or from the direct contact of airborne formaldehyde with the eyes or other tissue, and *systemic* effects are defined as effects that occur outside those systems.

[3]Ritchie, I.M., and R.G. Lehnen. 1987. Formaldehyde-related health complaints of residents living in mobile and conventional homes. Am. J. Public Health 77(3):323-328.

thermore, EPA set aside the chamber and occupational studies too soon in the process. Although the chamber studies are of acute duration, they are complementary with the residential studies and provide controlled measures of exposure and response. Therefore, the committee recommends that EPA present the concentration-response data from the occupational, chamber, and residential studies on the same graph and include the point estimate and measures of variability in the exposure concentrations and responses. The committee notes that EPA did not (but should) review research findings on transient-receptor-potential ion channels and evaluate the utility of this evidence for improving understanding of the mode of action for sensory irritation and respiratory effects attributed to formaldehyde exposure.

*Decreased pulmonary function.* The committee agrees with EPA that formaldehyde exposure may cause a decrease in pulmonary function, but EPA should provide a clear rationale to support that conclusion. Furthermore, although the committee supports the use of the study by Kryzanowski et al. (1990)[4] to calculate a candidate RfC, EPA should provide a clear description of how the study was used to estimate a point of departure and should also consider the studies conducted by Kriebel et al. (1993, 2001)[5] and the chamber studies for possible derivation of candidate RfCs.

*Respiratory tract pathology.* Animal studies in mice, rats, and nonhuman primates clearly show that inhaled formaldehyde at 2 ppm or greater causes cytotoxicity that increases epithelial-cell proliferation and that after prolonged inhalation can lead to nasal tumors. Although the committee agrees with EPA that the human studies that assessed upper respiratory tract pathology were insufficient to derive a candidate RfC, it disagrees with EPA's decision not to use the animal data. The animal studies offer one of the most extensive datasets on an inhaled chemical, and EPA should use the data to derive a candidate RfC for this outcome.

*Asthma.* Asthma is a term applied to a broad phenotype of respiratory disease that comprises an array of symptoms resulting from underlying airway inflammation and associated airway hyper-reactivity. In infants and children, wheezing illnesses that are the result of lower respiratory tract infections are often labeled as asthma, and in adults, the symptoms can overlap with those of other chronic diseases, such as chronic obstructive pulmonary disease. Thus, a critical review of the literature is essential to ensure that what is being evaluated is asthma. The committee notes that this issue is not adequately addressed in the

---

[4]Krzyzanowski, M., J.J. Quackenboss, and M.D. Lebowitz. 1990. Chronic respiratory effects of indoor formaldehyde exposure. Environ. Res. 52(2):117-125.

[5]Kriebel, D., S.R. Sama, and B. Cocanour. 1993. Reversible pulmonary responses to formaldehyde. A study of clinical anatomy students. Am. Rev. Respir. Dis. 148(6 Pt 1):1509-1515.

Kriebel, D., D. Myers, M. Cheng, S. Woskie, and B. Cocanour. 2001. Short-term effects of formaldehyde on peak expiratory flow and irritant symptoms. Arch. Environ. Health 56(1):11-18.

draft IRIS assessment and that EPA advanced a study (Rumchev et al. 2002)[6] that most likely suffers from misclassification of infection-associated wheezing in young children as asthma. The draft IRIS assessment also provides little discussion of the current understanding of the mechanisms of asthma causation and exacerbation. Given the abundant research available, the committee recommends that EPA strengthen its discussion of asthma to reflect current understanding of this complex disease and its pathogenesis. Although the committee agrees that the study by Garrett et al. (1999)[7] should be used to calculate a candidate RfC, the approach taken to identifying the point of departure needs further justification.

*Respiratory tract cancers.* The respiratory tract is considered to be a plausible location of formaldehyde-induced cancers in humans because these cancers occur at the site of first contact and because studies have shown an increased incidence of nasal tumors in rats and mice exposed to formaldehyde. However, the draft IRIS assessment does not present a clear framework for causal determinations and presents several conflicting statements that need to be resolved regarding the evidence of a causal association between formaldehyde and respiratory tract cancers. On the basis of EPA cancer guidelines, the committee agrees that there is sufficient evidence (that is, the combined weight of epidemiologic findings, results of animal studies, and mechanistic data) of a causal association between formaldehyde and cancers of the nose, nasal cavity, and nasopharynx. It disagrees that the evidence regarding other sites in the respiratory tract is sufficient. The committee agrees with EPA that the study by Hauptmann et al. (2004)[8] is the most appropriate for deriving a unit risk value but notes that this study is being updated.

## Systemic Health Effects

The systemic effects evaluated by EPA include immunotoxicity, neurotoxicity, reproductive and developmental toxicity, and lymphohematopoietic (LHP) cancers. As noted above, high reactivity and extensive nasal absorption of formaldehyde restrict systemic delivery of inhaled formaldehyde beyond the upper respiratory tract and major conducting airways of the lung, so systemic responses are unlikely to arise from the direct delivery of formaldehyde (or its hydrated form, methanediol) to a distant site in the body. However, a distinction

---

[6]Rumchev, K.B., J.T. Spickett, M.K. Bulsara, M.R. Phillips, and S.M. Stick. 2002. Domestic exposure to formaldehyde significantly increases the risk of asthma in young children. Eur. Respir. J. 20(2):403-408.

[7]Garrett, M.H., M.A. Hooper, B.M. Hooper, P.R. Rayment, and M.J. Abramson. 1999. Increased risk of allergy in children due to formaldehyde exposure in homes. Allergy 54(4):330-337 [Erratum-Allergy 54(12):1327].

[8]Hauptmann, M., J.H. Lubin, P.A. Stewart, R.B. Hayes, and A. Blair. 2004. Mortality from solid cancers among workers in formaldehyde industries. Am. J. Epidemiol. 159(12):1117-1130.

needs to be made between systemic delivery and systemic effects. The possibility remains that systemic delivery of formaldehyde is not a prerequisite for some of the reported systemic effects seen after formaldehyde exposure. Those effects may result from indirect modes of action associated with local effects, such as irritation, inflammation, and stress. Therefore, the committee reviewed EPA's evaluation of the systemic effects and determined whether the evidence presented supported EPA's conclusions.

As in the evaluation of the portal-of-entry effects, the committee concluded that EPA identified relevant literature and adequately described the studies selected; however, critical evaluations of study strengths and weaknesses were generally lacking, and clear rationales for conclusions were often not provided. As a result, some narratives did not support the conclusions stated. Comments on the specific outcomes are provided below.

*Immunotoxicity*. The draft IRIS assessment presents numerous studies suggesting that formaldehyde has the ability to affect immune functions. However, EPA should conduct a more rigorous evaluation of the strengths and weaknesses of the studies; more integration of the human and animal data would lend support to the conclusions made. The committee agrees with EPA's decision not to calculate a candidate RfC on the basis of immunotoxicity studies.

*Neurotoxicity*. The committee found that EPA overstated the evidence in concluding that formaldehyde is neurotoxic; the human data are insufficient, and the candidate animal studies deviate substantially from neurotoxicity-testing guidelines and common practice. Furthermore, the committee does not support EPA's conclusion that the behavioral changes observed in animals exposed to formaldehyde are not likely to be caused by the irritant properties of formaldehyde. Data indicate that those changes could occur as a result of nasal irritation or other local responses; stress, also an important confounder that can affect the nervous system, was not considered by EPA. The draft IRIS assessment provides conflicting statements that need to be resolved about whether formaldehyde is a direct neurotoxicant. The committee agrees with EPA's decision not to calculate a candidate RfC on the basis of the neurotoxicity studies.

*Reproductive and developmental toxicity*. The draft IRIS assessment states that epidemiologic studies provide evidence of a "convincing relationship between occupational exposure to formaldehyde and adverse reproductive outcomes in women." The committee disagrees and concludes that a small number of studies indicate a suggestive pattern of association rather than a "convincing relationship." Animal data also suggest an effect, but EPA should weigh the negative and positive results rigorously inasmuch as negative results outnumbered positive ones for some end points, should evaluate study quality critically because some studies of questionable quality were used to support conclusions, and should consider carefully potential confounders, such as maternal toxicity, effects of stress, exposure concentrations above the odor threshold, and potential for oral exposures through licking. Although the epidemiologic studies provide only a suggestive pattern of association, EPA followed its guidelines and chose the best available study to calculate a candidate RfC.

*Lymphohematopoietic cancers*. EPA evaluated the evidence of a causal relationship between formaldehyde exposure and several groupings of LHP cancers—"all LHP cancers," "all leukemias," and "myeloid leukemias." The committee does not support the grouping of "all LHP cancers" because it combines many diverse cancers that are not closely related in etiology and cells of origin. The committee recommends that EPA focus on the most specific diagnoses available in the epidemiologic data, such as acute myeloblastic leukemia, chronic lymphocytic leukemia, and specific lymphomas.

As with the respiratory tract cancers, the draft IRIS assessment does not provide a clear framework for causal determinations. As a result, the conclusions appear to be based on a subjective view of the overall data, and the absence of a causal framework for these cancers is particularly problematic given the inconsistencies in the epidemiologic data, the weak animal data, and the lack of mechanistic data. Although EPA provided an exhaustive description of the studies and speculated extensively on possible modes of action, the causal determinations are not supported by the narrative provided in the draft IRIS assessment. Accordingly, the committee recommends that EPA revisit arguments that support determinations of causality for specific LHP cancers and in so doing include detailed descriptions of the criteria that were used to weigh evidence and assess causality. That will add needed transparency and validity to its conclusions.

## Derivation of Reference Concentrations for Formaldehyde

An RfC is defined by EPA as "an estimate...of a continuous inhalation exposure to the human population...that is likely to be without an appreciable risk of deleterious effects during a lifetime" (EPA 2010).[9] It is derived by applying uncertainty factors to a point of departure that is identified in or derived from a study that evaluates a relevant health end point, such as asthma incidence. The committee was asked to comment on specific uncertainty factors used to derive the candidate RfCs in the draft IRIS assessment: the one used to capture variability in response to formaldehyde exposure in the human population ($UF_H$) and the one used to capture the adequacy of the database ($UF_D$). The committee notes that it had some difficulty in commenting on derivation of the RfCs because it would have made some different decisions regarding study selection and calculation of candidate RfCs as indicated above. Accordingly, the committee's comments here should not be interpreted as a recommendation for any particular RfC as presented in the draft IRIS assessment.

Determining the appropriate value of the $UF_H$ involves consideration of possible susceptibility of the human population and the factors that could influ-

---

[9]EPA (U.S. Environmental Protection Agency). 2010. Glossary, EPA Risk Assessment. U.S. Environmental Protection Agency [online]. Available: http://www.epa.gov/risk_assessment/glossary.htm#r [accessed Nov. 29, 2010].

ence it. The committee agrees with EPA that available data indicate that there are possible differences in susceptibility to formaldehyde at various life stages and in various disease states. The epidemiologic studies used to calculate the candidate RfCs for respiratory effects and sensory irritation included people in susceptible populations (children and people who have asthma). However, the modes of action for formaldehyde's effects are not sufficiently understood to ensure that all potential susceptible populations and factors contributing to susceptibility have been identified and adequately described. Thus, the committee supports the use of a $UF_H$ of 3 to calculate candidate RfCs for studies identified in the draft IRIS assessment on reduced pulmonary function, asthma, and sensory irritation, noting that the committee does not support the advancement of the studies by Richie and Lehnen (1987)[10] and Rumchev et al. (2002).[11]

Determining the appropriate value of the $UF_D$ involves consideration of the breadth and depth of the data available on a specific chemical. The database on formaldehyde is extensive and includes the evaluation of a full array of health outcomes in the human population and laboratory animals. Although there are some gaps in the data on reproductive, developmental, immunologic, and neurotoxic effects, the likelihood that new effects will be observed at concentrations below those at which respiratory effects have been observed is low. Thus, the committee supports the use of a $UF_D$ of 1 with the caveat that research of the types noted should be pursued.

Overall, the committee is troubled by the presentation and derivation of the proposed RfC values and strongly recommends the approach illustrated and described in Figure S-1. A similar approach was recommended by the NRC Committee to Review EPA's Toxicological Assessment of Tetrachloroethylene and used in recent EPA assessments of tetrachloroethylene and trichloroethylene. Appropriate graphic aids that enable the visualization of the concentration ranges of the candidate RfCs may identify a central value, isolate especially low or high RfC values that might not be consistent with the body of literature, and ultimately improve the ability of the assessment to make a compelling case that the RfC proposed is appropriate for the most sensitive end point and protective with regard to other potential health effects.

### Derivation of Unit Risk Estimates for Formaldehyde

Unit risk for formaldehyde can be defined as the estimate of extra risk caused by inhalation of one unit of formaldehyde, such as 1 ppm or 1 µg/m³, in

---

[10]Ritchie, I.M., and R.G. Lehnen. 1987. Formaldehyde-related health complaints of residents living in mobile and conventional homes. Am. J. Public Health 77(3):323-328.

[11]Rumchev, K.B., J.T. Spickett, M.K. Bulsara, M.R. Phillips, and S.M. Stick. 2002. Domestic exposure to formaldehyde significantly increases the risk of asthma in young children. Eur. Respir. J. 20(2):403-408.

# Summary

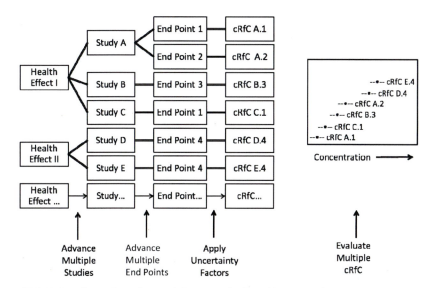

**FIGURE S-1** Illustration of potential process for identifying an RfC. Health effects associated with exposure to the chemical are identified. For each health effect, studies that meet inclusion criteria are advanced. From each study, one or more health end points that meet specified criteria are advanced, and a point of departure is identified or derived. Uncertainty factors are selected and applied to the point of departure to yield a candidate RfC (cRfC). All cRfCs are evaluated together with the aid of graphic displays that incorporate selected information relevant to the database and to the decision to be made. A final RfC is selected from the distribution after consideration of all critical data that meet the inclusion criteria.

air. EPA used studies of the National Cancer Institute (NCI) cohort of U.S. workers exposed to formaldehyde through its production or its use (Hauptmann et al. 2004[12]; Beane-Freeman et al. 2009[13]) to estimate unit risk values for three cancers—nasopharyngeal cancer, Hodgkin lymphoma, and leukemia. The committee agrees that the NCI studies are a reasonable choice because they are the only ones with exposure and dose-response data sufficient for calculation of the unit risks; however, the studies are not without their weaknesses, which should be clearly discussed and addressed in the revised IRIS assessment. Although there are uncertainties as discussed above regarding the causal relationship of

---

[12] Hauptmann, M., J.H. Lubin, P.A. Stewart, R.B. Hayes, and A. Blair. 2004. Mortality from solid cancers among workers in formaldehyde industries. Am. J. Epidemiol. 159(12):1117-1130.

[13] Beane-Freeman, L.E., A. Blair, J.H. Lubin, P.A. Stewart, R.B. Hayes, R.N. Hoover, and M. Hauptmann. 2009. Mortality from lymphohematopoietic malignancies among workers in formaldehyde industries: The National Cancer Institute cohort. J. Natl. Cancer Inst. 101(10):751-761.

formaldehyde exposure and the three kinds of cancer, EPA's decision to calculate unit risk values for them appears to be defensible on the basis of the agency's cancer guidelines. However, EPA should provide a clear description of the criteria that it used to select the specific cancers and demonstrate a systematic application of the criteria. The calculation of the unit risk values is a complex process, involves many sources of uncertainty and variability, and is influenced by the low-dose extrapolation used (for example, linear vs threshold). The committee therefore recommends that EPA conduct an independent analysis of the dose-response models to confirm the degree to which the models fit the data appropriately. EPA is encouraged to consider the use of alternative extrapolation models for the analysis of the cancer data; this is especially important given the use of a single study, the inconsistencies in the exposure measures, and the uncertainties associated with the selected cancers.

## THE FORMALDEHYDE IRIS ASSESSMENT: THE PATH FORWARD

The committee recognizes that the completion of the formaldehyde IRIS assessment is awaited by diverse stakeholders, and it has tried to be judicious in its recommendations of specific changes noted in its report. However, the committee concludes that the following general recommendations are critical to address in the revision of the draft assessment. First, rigorous editing is needed to reduce the volume of the text substantially and address the redundancies and inconsistencies; reducing the text could greatly enhance the clarity of the document. Second, Chapter 1 of the draft assessment needs to discuss more fully the methods of the assessment. The committee is recommending not the addition of long descriptions of EPA guidelines but rather clear concise statements of criteria used to exclude, include, and advance studies for derivation of the RfCs and unit risk estimates. Third, standardized evidence tables that provide the methods and results of each study are needed for all health outcomes; if appropriate tables were used, long descriptions of the studies could be moved to an appendix or deleted. Fourth, all critical studies need to be thoroughly evaluated for strengths and weaknesses by using uniform approaches; the findings of these evaluations could be summarized in tables to ensure transparency. Fifth, the rationales for selection of studies that are used to calculate RfCs and unit risks need to be articulated clearly. Sixth, the weight-of-evidence descriptions need to indicate the various determinants of "weight." The reader needs to be able to understand what elements (such as consistency) were emphasized in synthesizing the evidence.

The committee is concerned about the persistence of problems encountered with IRIS assessments over the years, especially given the multiple groups that have highlighted them, and encourages EPA to address the problems with development of the draft assessments that have been identified. The committee recognizes that revision of the approach will involve an extensive effort by EPA staff and others, and it is not recommending that EPA delay the revision of the

*Summary*

formaldehyde assessment to implement a new approach. However, models for conducting IRIS assessments more effectively and efficiently are available, and the committee provides several examples in the present report. Thus, EPA might be able to make changes in its process relatively quickly by selecting and adapting existing approaches. As exemplified by the recent revision of the approach used for the National Ambient Air Quality Standards, this task is not insurmountable. If the methodologic issues are not addressed, future assessments may still have the same general and avoidable problems that are highlighted here.

# 1

# Introduction

Health effects from exposure to formaldehyde have been a topic of research for decades. Past concerns arose from exposures in indoor environments and studies of workers showing increased risk of nasopharyngeal cancer. In recent years, people who were displaced by Hurricane Katrina and Hurricane Rita and lived in trailers provided by the Federal Emergency Management Agency have reported adverse health effects attributed to formaldehyde exposure. Published research has also indicated a possible link between leukemia and formaldehyde exposure.

The U.S. Environmental Protection Agency (EPA) has been working to update its health assessment of formaldehyde for its Integrated Risk Information System (IRIS) for a number of years. The large amount of new research and data on formaldehyde since its original assessment in the early 1990s has made the task challenging. Given the complex nature of the assessment and the knowledge that the assessment will be used as the basis of regulatory decisions, EPA asked the National Research Council (NRC) to conduct an independent scientific review of the draft IRIS assessment and answer questions related specifically to its derivation of reference concentrations (RfCs) for noncancer effects and of its unit risk estimates for cancer. In response to EPA's request, NRC convened the Committee to Review EPA's Draft IRIS Assessment of Formaldehyde, which prepared this report.

## FORMALDEHYDE AND THE DRAFT ASSESSMENT

Formaldehyde, which has the chemical structure shown in Figure 1-1, is a chemical building block of numerous compounds that are used in a wide array of products (see Gerberich and Seaman 1994; ATSDR 1999; IARC 2006). One main use is to make resins that are used as adhesives in the production of particle board, fiberboard, plywood, and other wood products. The resins are also used to make molding and insulating materials and are used in a variety of other industries, including the textile, rubber, and cement industries.

# Introduction

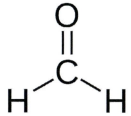

**FIGURE 1-1** Formaldehyde chemical structure. Formaldehyde is described as a colorless gas at room temperature with a pungent, suffocating odor.

Formaldehyde is a common environmental chemical that is found in ambient and indoor air. It is also present naturally in some foods and is a metabolic intermediate in the human body. For ambient air, major emission sources include power plants, incinerators, refineries, manufacturing facilities, and automobiles (ATSDR 1999; IARC 2006). Formaldehyde is also produced by vegetative decay, animal wastes, forest fires, and photochemical oxidation of hydrocarbons in the lower atmosphere (ATSDR 1999; IARC 2006). The most recent EPA data on ambient-air concentrations indicate that the annual means at monitoring sites range from 0.56 to 36.31 ppb, and the overall mean is 2.77 ppb (EPA 2010). If the data are categorized by land use, agricultural locations have the lowest mean, 1.68 ppb, and locations affected primarily by mobile sources have the highest mean, 5.52 ppb.

Indoor air typically has higher formaldehyde concentrations than ambient air (ATSDR 1999; IARC 2006; EPA 2010). Major indoor emission sources include building materials, consumer products, gas and wood stoves, kerosene heaters, and cigarettes. Indoor-air concentrations depend on the age and type of construction. Older conventional homes have lower formaldehyde concentrations than newer constructions, and conventional homes have lower formaldehyde concentrations than mobile homes. Formaldehyde concentrations in indoor air have been decreasing since the 1980s, when restrictions on formaldehyde emissions from building materials were tightened (ATSDR 1999; EPA 2010; Salthammer et al. 2010). However, on the basis of a review of international studies, Salthammer et al. (2010) estimated the average formaldehyde exposure of the general population to be 16-32 ppb in air. Figure 1-2 provides ranges of formaldehyde air concentrations in various environments.

Given the pervasive exposure of the general population to some concentration of formaldehyde, federal agencies tasked with protecting public health are concerned about the health effects of formaldehyde exposure. EPA is re-evaluating regulations on the emissions of formaldehyde from composite wood products and, as part of that effort, is re-evaluating its assessment of noncancer and cancer risks associated with formaldehyde. Figure 1-3 provides a timeline of EPA's activity since its original assessments of noncancer and cancer risks were released in 1990 and 1991, respectively.

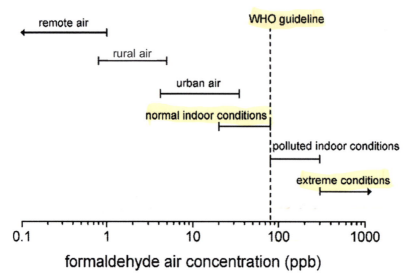

**FIGURE 1-2** Formaldehyde concentration in various environments. Abbreviation: WHO, World Health Organization. Source: Salthammer et al. 2010. Reprinted with permission; copyright 2010, American Chemical Society.

Since 1991, numerous studies of the toxicity and carcinogenic potential of formaldehyde have been published. In 2006, the International Agency on Cancer Research (IARC) revised its formaldehyde classification from probably carcinogenic to humans (Group 2A) to carcinogenic to humans (Group 1). The revision was based on what IARC concluded to be sufficient evidence of nasopharyngeal cancer in humans, strong but not sufficient evidence of leukemia in humans, and limited evidence of sinonasal cancer in humans (IARC 2006). In 2009, IARC reaffirmed its classification of formaldehyde but concluded that there was sufficient evidence of leukemia in humans (Baan et al. 2009). Furthermore, in 2010, an expert National Toxicology Program (NTP) panel on formaldehyde recommended that formaldehyde be listed as a known human carcinogen in its *Report on Carcinogens* (McMartin et al. 2009). That recommendation was a change from the previous edition, which listed formaldehyde as "reasonably anticipated to be a human carcinogen" (NTP 2005). Some scientists do not agree with the recent conclusions from IARC and NTP and have published new studies that they claim cast doubt on them.

Given the complex nature of assessing the health effects of formaldehyde and the knowledge that the IRIS assessment will be used as a basis of new regulations, EPA asked NRC to convene a committee to review its draft IRIS assessment.

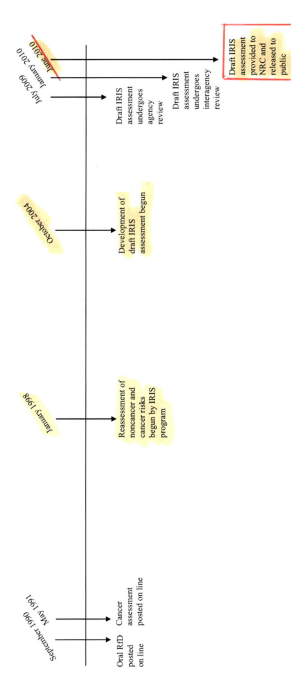

**FIGURE 1-3** Timeline of the development of the draft IRIS assessment. Abbreviations: IRIS, Integrated Risk Information System; NRC, National Research Council; RfD, and reference dose.

## THE COMMITTEE'S TASK AND APPROACH

The committee convened as a result of EPA's request included experts in epidemiology, exposure assessment, leukemogenesis, mechanisms of carcinogenicity, inhalation toxicology, neurotoxicology, reproductive and developmental toxicology, statistics, physiologically based pharmacokinetic modeling, and risk assessment (see Appendix A for biographic information on the committee). The committee was asked to review EPA's draft IRIS assessment and to answer questions concerning the identification of potential noncancer health effects, the selection of the points of departure for those health effects, and the basis of the determination of uncertainty factors used to derive the RfCs. The committee was also asked specifically to comment on the scientific rationale provided for the cancer assessment and the quantified estimates derived. The verbatim statement of task is provided in Box 1-1.

---

**BOX 1-1** Statement of Task

A committee of the National Research Council (NRC) will conduct an independent scientific review of the U.S. Environmental Protection Agency (EPA) draft human health assessment of formaldehyde for the Integrated Risk Information System (IRIS). The committee will provide a brief report that comments on EPA's identification of potential adverse noncancer health effects, assessment of carcinogenic potential, exposure-response analysis for identified end points, quantitative risk assessment methods, and evaluation of sources of uncertainty in the health assessment. Specifically, the committee will address tasks such as the following:

*Inhalation Reference Concentration for Formaldehyde*

- Review and comment on the draft's analysis of the potential noncancer health effects attributable to inhalation exposure to formaldehyde and answer the following questions: Has EPA fairly and soundly evaluated the weight of evidence that formaldehyde causes the effects identified in the assessment? Has it reached conclusions that can be supported by the available studies and appropriately identified and described the weaknesses of the studies?
- Review and comment on the draft's evaluation of the studies used to identify the points of departure for quantitative derivation of the reference concentration and answer the following questions: Has EPA selected studies of suitable quality for the quantitative analysis? Has it appropriately determined the points of departure for the effects? In addition, review and comment on EPA's determinations as to when and how to adjust appropriately for exposure duration and whether alternatives were adequately considered and presented.

*(Continued)*

# Introduction
21

> **BOX 1-1** Continued
>
> • Review and comment on the draft's evaluation of the studies used to determine the uncertainty factors for derivation of the reference concentration for the sensitive noncancer effects of formaldehyde. Also, review and discuss the evaluation of the extent to which the available studies capture the range of human variability in response to formaldehyde exposures; and review and discuss the completeness of the database used to identify the hazards of formaldehyde inhalation and to derive a reference concentration.
>
> *Carcinogenicity of Formaldehyde*
>
> • Comment on the cancer weight-of-evidence narrative in the draft, developed according to EPA's 2005 Guidelines for Carcinogen Risk Assessment and answer the question, is the weight-of-evidence narrative scientifically supported?
> • Review and comment on the draft's reasonable upper estimates of the potential human cancer risk attributable to inhalation of formaldehyde at low concentrations.
> • Review and comment on the scientific support for the choices made in developing the preferred quantitative estimates that are based on dose-response relationships between several cancers and cumulative inhalation exposure, and consider such issues as the appropriate dose metric given the study design, the alternative metrics, and the suitability of alternative metrics for use in evaluating environmental and residential inhalation exposures to formaldehyde.
> • Review and comment on the scientific rationale for the choices made to develop the supportive estimates that are based on dose-response relationships from animal studies of nasal tumors, and consider the analysis of the sensitivity of low-dose estimates from biologically based dose-response models of formaldehyde for upper respiratory tract cancer to small changes in model design or model inputs.

To accomplish its task, the committee held four meetings from June 2010 to December 2010. The first two meetings included public sessions during which the committee heard primarily from the sponsor on the development of the draft IRIS assessment and approaches used to derive the estimates presented in it. During each public session, interested parties addressed the committee. The committee reviewed the draft assessment, numerous scientific publications, and all materials submitted to it by outside parties.

The committee was tasked with conducting an "independent scientific review" of the draft IRIS assessment, not with conducting its own assessment. Therefore, the committee did not conduct its own literature search, review all relevant evidence, systematically formulate its own conclusions regarding causality, or recommend values for the RfC and unit risk. The committee reviewed the draft IRIS assessment and its methods and key literature and determined whether EPA's conclusions were supported on the basis of that assessment and literature. Thus, the present report contains the committee's conclusions and recommendations resulting from its review of the draft assessment. The committee notes that it does not provide a comprehensive discussion of any particular topic or health outcome, although it does provide brief descriptions where necessary to give the reader some context as to what it is recommending. Furthermore, the committee discussed the various health outcomes using the categories presented in the draft IRIS assessment. Some overlap among the categories was noted; for example, asthma—a disease with an immunologic basis—was handled separately from immunologic effects.

Because the committee evaluated what EPA did, there is some inherent variability in the depth of the committee's review given the varied discussions in the draft IRIS assessment. For example, the draft assessment presents discussions on mode of action that vary in level of detail, analysis, and referencing. In some cases, mode-of-action data—which would support EPA's conclusion—are available, but they are not presented in the draft assessment. In those cases, the committee recommends that those data be reviewed and evaluated. In other cases, mode of action is highly speculative, and the speculations are discussed at length in the draft assessment. In those cases, the committee recommends that the discussion be truncated given the speculative nature of the hypotheses. The committee notes that a well-established mode of action is not required to make causal inferences, but mode-of-action data should be discussed when those data support EPA's conclusions.

## ORGANIZATION OF REPORT

The committee organized its report by separating the overarching elements of its charge from the more specific ones. Specifically, Chapter 2 addresses the general methods to develop the draft IRIS assessment because the committee has concerns about the methods used in its development. Chapter 3 reviews the toxicokinetics of formaldehyde, which has general relevance for effects at the portal of entry and elsewhere, and therefore this review precedes the other chapters. The remaining chapters were structured to address the specific elements of the charge related to the RfCs and unit risk. Accordingly, Chapters 4 and 5 discuss the weight of evidence for hazard identification and study selection for portal-of-entry and systemic effects, respectively, and Chapter 6 addresses the derivation of the RfC and unit risk. Chapter 7 provides general recommendations for revisions of the draft assessment and, on the basis of the findings in Chapters 2-6, comments on the IRIS process used to generate the present assessment.

## REFERENCES

ATSDR (Agency for Toxic Substances and Disease Registry). 1999. Toxicological Profile for Formaldehyde. U.S. Department of Health and Human Services, Public Health Services, Agency for Toxic Substances and Disease Registry, Atlanta, GA. July 1999 [online]. Available: http://www.atsdr.cdc.gov/ToxProfiles/tp111.pdf [accessed Jan. 5, 2011].

Baan, R., Y. Grosse, K. Straif, B. Secretan, F. El Ghissassi, V. Bouvard, L. Benbrahim-Tallaa, N. Guha, C. Freeman, L. Galichet, and V. Cogliano. 2009. A review of human carcinogens—Part F: Chemical agents and related occupations. Lancet Oncol. 10(12):1143-1144.

EPA (U.S. Environmental Protection Agency). 2010. Toxicological Review of Formaldehyde (CAS No. 50-00-0) – Inhalation Assessment: In Support of Summary Information on the Integrated Risk Information System (IRIS). External Review Draft. EPA/635/R-10/002A. U.S. Environmental Protection Agency, Washington, DC [online]. Available: http://cfpub.epa.gov/ncea/iris_drafts/recordisplay.cfm?deid=223614 [accessed Nov. 22, 2010].

Gerberich, H.R., and G.C. Seaman. 1994. Formaldehyde. Pp. 929-951 in Kirk-Othmer Encyclopedia of Chemical Technology, Vol. 11, 4th Ed., J.I. Kroschwitz, and M. Howe-Grant, eds. New York: Wiley.

IARC (International Agency for Research on Cancer). 2006. Formaldehyde. Pp. 39-325 in Formaldehyde, 2-Butoxyethanol and 1-tert-Butoxypropan-2-ol. IARC Monographs on the Evaluation of Carcinogenic Risks to Humans, Vol. 88. Lyon, France: International Agency for Research on Cancer.

McMartin, K.E., F. Akbar-Khanzadeh, G.A. Boorman, A. DeRoos, P. Demers, L. Peterson, S. Rappaport, D.B. Richardson, W.T. Sanderson, M.S. Sandy, L.B. Freeman, M. DeVito, S.A. Elmore, and L. Zhang. 2009. Part B—Recommendations for the Listing Status for Formaldehyde and Scientific Justification for the Recommendation. Formaldehyde Expert Panel Report Part B [online]. Available: http://ntp.niehs.nih.gov/ntp/roc/twelfth/2009/November/FA_PartB.pdf [accessed Jan. 5, 2011].

NTP (National Toxicology Program). 2005. Formaldehyde (Gas): CAS No. 50-00-0. Substance Profiles. Report on Carcinogens, 11 Ed. U.S. Department of health and Human Services, Public Health Service, National Toxicology Program [online]. Available: http://ntp.niehs.nih.gov/ntp/roc/eleventh/profiles/s089form.pdf [accessed Jan. 5, 2011].

Salthammer, T., S. Mentese, and R. Marutzky. 2010. Formaldehyde in the indoor environment. Chem. Rev. 110(4):2536-2572.

# 2

# Review of Methods

As noted in Chapter 1, the committee was asked to review and comment on specific aspects of the draft IRIS assessment of formaldehyde. This chapter provides general comments on the methods and structure of the document. The committee's rationale for providing general comments is that the specific elements of the charge are inseparable from the approach used by the U.S. Environmental Protection Agency (EPA) for the development of the assessment and presentation of its findings.[1] In responding to questions posed in its charge and developing its report, the committee noted some recurring methodologic problems that cut across components of the charge.

The general problems that the committee identified are not unique to the draft IRIS assessment of formaldehyde. Committees of the Board on Environmental Studies and Toxicology (BEST) of the National Research Council (NRC) have reviewed a number of IRIS assessments in the last decade, including three (NRC 2005, 2006, 2010) in the last 5 years. Some of the general problems identified by the present committee have been commented on by the other BEST committees. For example, the 2006 NRC report on dioxin and related compounds commented on the need for formal, evidence-based approaches for noncancer effects, the need for transparency and clarity in the selection of data sets for analysis, and the need for greater attention to uncertainty and variability (NRC 2006). The 2010 NRC review of the draft IRIS assessment of tetrachloroethylene found similar problems and provided a chapter, "Moving Beyond the Current State of Practice," that addressed methodologic issues and the failure to establish clear and transparent methods for carrying out and presenting the assessment (NRC 2010). That report also provided a broad set of recommendations on characterization of uncertainty.

---

[1]The committee distinguishes between the process used to generate the draft IRIS assessment and the overall IRIS process that includes not only generation of the assessment but the many layers of review. In this report, the committee is focused on the approach used to generate the draft assessment.

*Review of Methods* 25

The present chapter addresses the general assessment methods and covers identification of the studies considered, their evaluation, and the weight-of-evidence assessment. These issues are also addressed within the context of the specific health outcomes evaluated (see Chapters 4-5).

## REVIEW OF THE METHODOLOGY OF THE DRAFT IRIS ASSESSMENT OF FORMALDEHYDE

IRIS has the overall purpose of evaluating human health effects that may arise from exposure to environmental contaminants (EPA 2010a). An IRIS assessment addresses noncancer and cancer effects as appropriate and provides descriptive and quantitative information:

• "Noncancer effects: Oral reference doses and inhalation reference concentrations (RfDs and RfCs, respectively) for effects known or assumed to be produced through a nonlinear (possibly threshold) mode of action. In most instances, RfDs and RfCs are developed for the noncarcinogenic effects of substances" (EPA 2010a).

• "Cancer effects: Descriptors that characterize the weight of evidence for human carcinogenicity, oral slope factors, and oral and inhalation unit risks for carcinogenic effects. Where a nonlinear mode of action is established, RfD and RfC values may be used" (EPA 2010a).

A sequence of activities is involved in conducting IRIS assessments and in calculating RfCs and unit risk estimates. Figure 2-1 is a generic schema that describes the steps used to generate the draft IRIS assessment and the actions needed at each step. The figure is the committee's representation of that process, as gleaned from the assessment. Although the draft IRIS assessment does not explicitly acknowledge these steps, they are implicit in the approach and are ordered as shown.

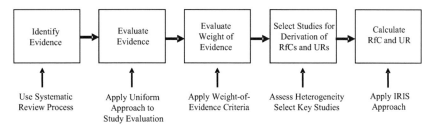

**FIGURE 2-1** Elements of the IRIS process. Abbreviations: IRIS, Integrated Risk Information System; RfC, reference concentration; and UR, unit risk.

In fact, the general approach receives little description in the draft assessment. The methods for conducting the assessment are described in a two-page introduction to a document whose main chapters and appendixes occupy about 1,000 pages. The committee notes that the introductory chapter of the formaldehyde assessment is almost identical with that of other IRIS assessments (see, for example, the IRIS summary for chloroprene, EPA 2010b). The two pages constitute the sole description of the methods used by the authors and cite numerous EPA guidance documents, some dating to 1986 (see Appendix B for some of the most relevant portions of those documents). Some of the guidance documents are cited at appropriate points in the draft assessment, but their specific roles in the preparation of the draft are not clear.

In general, the committee found that the draft assessment was not prepared in a transparent, consistent fashion with clear linkages to an underlying framework as it moves from review of the relevant evidence to calculation of the RfCs and unit risk estimates and characterization of their uncertainty and variability. The committee did not find sufficient documentation of methods and criteria for identifying the epidemiologic and experimental evidence to be reviewed, for evaluating individual studies, for assessing weight of evidence, for selecting individual studies for derivation of toxicity and risk estimates, or for characterizing uncertainty and variability. Summary sections that synthesize the evidence are variable and too often brief or not present, and strength of evidence is not characterized with standardized descriptors.

The committee emphasizes that its criticism regarding the lack of documentation is not a recommendation for adding lengthy summaries of the individual guidance documents to the introductory chapter. It is suggesting that clear concise descriptions of key criteria used to include studies in the analysis, to exclude studies, or to advance studies for calculation of RfCs and unit risk estimates are needed. Nuances concerning specific health outcomes could be addressed in the introductory sections on those outcomes. The following sections provide comments on the general steps of the process. Again, specific aspects are addressed in Chapters 4-5 for each health outcome.

## Literature Identification

The ability to identify and filter studies is crucial for any literature review that is synthesizing the potential effects of a suspected hazard. The evaluation of all relevant studies in an IRIS review process is analogous to the collection of relevant studies for a meta-analysis. A general approach to literature review is provided in Chapter 1 of the draft IRIS assessment of formaldehyde. EPA used a Chemical Abstracts Service Registry Number (CASRN) and "at least one common name" (EPA 2010c, p. 1-2) to search for relevant publications. The specific databases searched are not listed. PubMed searches are critical for identifying the literature on risks to health; the committee notes that PubMed does not specifically capture the CASRN.

The state of the art of literature searches now involves providing an extensive description of the databases searched and the search terms used. Chapter 1 of the draft IRIS assessment does not provide a list of the search terms used, such as terms that were used for the various health outcomes relevant to formaldehyde. The draft assessment also does not describe the results of searches, so the numbers of articles identified and excluded are unavailable to readers.

## Study Evaluation

The draft IRIS assessment evaluates many individual studies in a variety of disciplines. A description of the methods for evaluating individual studies is not provided, and it appears to this committee that studies were not reviewed with a common template for assessing their strengths and weaknesses. The committee notes that the template for evaluation would vary appropriately with the type of research study being considered. Such a strategy is not uniformly evident, and the evidence considered is not presented consistently in informative tables.

In some cases, there is a tendency to describe the studies ultimately selected for the derivation of the RfC in favorable terms. For example, a cross-sectional study by Krzyzanowski et al. (1990)—a study selected for the calculation of an RfC for respiratory effects—is referred to as "well-designed and executed" (EPA 2010c, p. 4-41) without emphasis on the inherent weaknesses of its cross-sectional design. The committee found one study selected for advancement for calculation of an RfC (Ritchie and Lehnen 1987) to be potentially subject to severe bias and would not have recommended it for advancement. Specifically, selection of the study population was based on a visit to a physician and referral for formaldehyde-concentration measurement, and the concentration-response gradient was considered by the committee to be implausibly steep (see Chapter 4 for further discussion).

## Synthesis of Evidence and Evaluation of Causation

In evaluating the evidence of causation, the draft IRIS assessment cites various EPA guidelines that apply weight-of-evidence approaches in assessing the strength of evidence. Those guidelines have been developed over a period of nearly 2 decades, and consequently consistency of methods is lacking from outcome to outcome (Appendix B). The implementation of the guidelines appears to be subjective and not standardized. The committee found variable detail in how the weight-of-evidence criteria had been applied. Uniformly developed discussions applying the weight-of-evidence criteria cannot be identified at appropriate points in the text. In some sections, the discussion of biologic evidence is particularly weak (for example, in the case of asthma pathogenesis) and often not reflective of the current state of knowledge.

## SUMMARY

In summary, when the review of studies used in the draft IRIS assessment of formaldehyde is compared with the current standard for evidence-based reviews and causal inference, limitations in each step used to generate the draft IRIS assessment are evident. For example, the methods are not clearly described, the review approaches are not transparent, and there is no indication that evidence-grading strategies were uniformly applied. In addition, the selection approach to identifying studies for RfC calculation appears ad hoc. The committee emphasizes that it is not recommending that EPA add an extensive discussion of its guidelines to the draft IRIS assessment. It is recommending that key factors used to exclude, include, or advance studies be discussed.

## REFERENCES

EPA (U.S. Environmental Protection Agency). 2010a. Frequent Questions: What is IRIS? Integrated Risk Information System, U.S. Environmental Protection Agency, Washington DC [online]. Available: http://www.epa.gov/IRIS/help_ques.htm#whatiris [accessed Dec. 28, 2010].

EPA (U.S. Environmental Protection Agency). 2010b. Toxicological Review of Chloroprene (CAS No. 126-99-8). In Support of Summary Information on the Integrated Risk Information System (IRIS). EPA/635/R-09/01F. U.S. Environmental Protection Agency, Washington, DC. September 2010 [online]. Available: http://www.epa.gov/iris/toxreviews/1021tr.pdf [accessed Jan. 6, 2010].

EPA (U.S. Environmental Protection Agency). 2010c. Toxicological Review of Formaldehyde (CAS No. 50-00-0) – Inhalation Assessment: In Support of Summary Information on the Integrated Risk Information System (IRIS). External Review Draft. EPA/635/R-10/002A. U.S. Environmental Protection Agency, Washington, DC [online]. Available: http://cfpub.epa.gov/ncea/iris_drafts/recordisplay.cfm?deid=223614 [accessed Nov. 22, 2010].

Krzyzanowski, M., J.J. Quackenboss, and M.D. Lebowitz. 1990. Chronic respiratory effects of indoor formaldehyde exposure. Environ. Res. 52(2):117-125.

NRC (National Research Council). 2005. Health Implications of Perchlorate Ingestion. Washington, DC: National Academies Press.

NRC (National Research Council). 2006. Health Risks from Dioxin and Related Compounds: Evaluation of the EPA Reassessment. Washington, DC: National Academies Press.

NRC (National Research Council). 2010. Review of the Environmental Protection Agency's Draft IRIS Assessment of Tetrachloroethylene. Washington, DC: National Academies Press.

Ritchie, I.M., and R.G. Lehnen. 1987. Formaldehyde-related health complaints of residents living in mobile and conventional homes. Am. J. Public Health. 77(3):323-328.

# 3

# Toxicokinetics and Modes of Action of Formaldehyde

This chapter provides the committee's review of the draft IRIS assessment that is relevant to formaldehyde toxicokinetics, carcinogenic modes of action, pharmacokinetic models, and biologically based dose-response (BBDR) models.[1] The committee comments on the Environmental Protection Agency (EPA) analysis of the fate of inhaled formaldehyde in the respiratory tract (portal of entry) and at more distant sites reached through systemic circulation, the use of formaldehyde-induced cross-links as biomarkers, and the ability of formaldehyde to cause systemic genotoxic effects. The committee also reviews EPA's use of the computational pharmacokinetic models and BBDR models that have been developed for formaldehyde and considers EPA's analysis of the sensitivity of low-dose BBDR-model estimates to small changes in model design or model inputs.

The discussion provided here is not intended to be exhaustive but rather focuses on the evidence presented in the draft IRIS assessment that was used to support EPA's key conclusions. It also dwells on the inhalation pathway rather than other exposure pathways because the inhalation pathway is the focus of the draft IRIS assessment. In conducting its review, the committee attempted to answer several central questions underlying the approach taken by EPA, including the following:

- Is formaldehyde an endogenous chemical?
- What is the immediate fate of inhaled formaldehyde?
- Is inhaled formaldehyde available systemically?

---

[1]This chapter focuses on carcinogenic modes of action. Known or hypothesized modes of action for other effects, such as airway irritation, are discussed elsewhere in this report.

- Can formaldehyde-related effects alter its toxicokinetics?
- Are formaldehyde-induced cross-links useful biomarkers of exposure?
- Can inhaled formaldehyde have systemic genotoxic effects?
- Are useful computational pharmacokinetic models for formaldehyde inhalation available?
  - Has a mode of action for formaldehyde carcinogenesis been identified?
  - What is the status of BBDR models for formaldehyde?
- Should the BBDR models available for formaldehyde be used in EPA's quantitative assessment?

Some of those questions could be answered by weighing the evidence from research studies considered in the draft IRIS assessment; others could not be answered by the committee with high confidence.

Overall, the committee found that the chapters describing the toxicokinetics, modes of action, and various models are well organized and that the draft IRIS assessment accurately reflects the current understanding of the toxicokinetics of inhaled formaldehyde and provides a thorough review of the metabolism, cytotoxicity, and genotoxicity of formaldehyde. The literature review in the draft IRIS assessment appears to be up to date and to include all major and recent studies published as of the release date.

## TOXICOKINETICS

### Is Formaldehyde an Endogenous Chemical?

The committee notes that EPA satisfactorily describes the current understanding of endogenous formaldehyde. It is well established that formaldehyde is produced endogenously by enzymatic and nonenzymatic pathways or as a detoxification product of xenobiotics during cellular metabolism (ATSDR 1999). There is also broad agreement that formaldehyde originating from metabolic or dietary sources is normally present at low concentrations in all tissues, cells, and bodily fluids. The concentration of endogenous formaldehyde in the blood of rats, monkeys, and humans is about 0.1 mM (Heck et al. 1985; Casanova et al. 1988). Background concentrations in the liver and nasal mucosa of the rat are 2-4 times those in the blood (Heck et al. 1982). Endogenous tissue formaldehyde concentrations are similar to concentrations (about 0.05 mM) that induce genotoxicity and cytolethality in vitro (Heck and Casanova 2004).

Heck et al. (1985) did not observe an increase in blood formaldehyde concentrations in rats and humans after exposure to inhaled formaldehyde at 14.4 ppm (2 hr) or 1.9 ppm (40 min), respectively. Subchronic studies conducted in rhesus monkeys have also shown that blood formaldehyde concentration was not measurably altered by exposure to airborne formaldehyde at 6 ppm for 6 hr/day 5 days/week for 4 weeks (Casanova-Schmitz et al. 1984).

Formaldehyde has also been measured in exhaled breath, but the interpretation of some measurements made with mass spectrometry has been questioned (Spanel and Smith 2008; Schripp et al. 2010). Spanel and Smith (2008) showed that a trace contaminant (up to 1%) of the reagent gas used in real-time mass-spectrometric methods—specifically proton-transfer reaction mass spectrometry (PTR-MS) and selected ion flow tube mass spectrometry (SIFT-MS)—reacts with endogenous methanol and ethanol that is normally found in exhaled breath to produce the same main ion (mass-to-charge ratio of 31) as is used to measure formaldehyde. Thus, they concluded that up to 5 ppb of the formaldehyde concentration determined in the exhaled breath of humans reported in earlier studies that did not account for this confounding may be due to methanol or ethanol and not formaldehyde; that is, 1% of total background concentrations of methanol or ethanol of about 500 ppb would be misclassified as formaldehyde. The committee concurs with EPA's concerns as to whether some published exhaled-breath measurements of formaldehyde are analytically valid. The committee also notes that this methodologic problem is inconsistently addressed by EPA in its reanalysis of the exhaled-breath experiments. The committee concludes, however, that regardless of the methodologic issue related to breath analysis, formaldehyde is normally present at a few parts per billion in exhaled breath after the measurement error associated with a trace contaminant in the reagent gas used in previous mass-spectrometric methods is taken into account.

The committee concludes that formaldehyde is an endogenous compound and that this finding complicates assessments of the risk posed by inhalation of formaldehyde. The committee emphasizes that the natural presence of various concentrations of formaldehyde in target tissues remains an important uncertainty with regard to assessment of the additional dose received by inhalation.

## What Is the Immediate Fate of Inhaled Formaldehyde?

Formaldehyde has been the subject of multiple toxicokinetic studies in rodents, dogs, and monkeys and of numerous in vitro and biomonitoring studies in humans. Although there may be quantitative differences between species, the fate of formaldehyde is qualitatively similar among species. The draft IRIS assessment provides an extensive and thorough review of the literature on the fate of formaldehyde in the body.

Formaldehyde is highly water-soluble and exists in water almost exclusively in a reversible hydrated form (methanediol). Thus, less than 0.1% of formaldehyde can be considered "free" once it enters the body. Formaldehyde is highly reactive at the site of entry and reacts readily with macromolecules, including DNA to form DNA-protein (DPX) and DNA-DNA (DDX) cross-links. Formaldehyde is oxidized to formate by a low-Km (400-μM) mitochondrial aldehyde dehydrogenase-2 (ALDH2) or via a two-enzyme system that converts nonenzymatically formed glutathione adducts (*S*-hydroxymethylglutathione) to the intermediate *S*-formylglutathione, which is then metabolized to formate and glutathione by *S*-formylglutathione hydrolase. Those metabolic processes contribute to the short biologic half-life of formaldehyde.

Inhaled formaldehyde is absorbed primarily in the upper airways because of its high water solubility, metabolism, and reactivity. Inhaled formaldehyde in the nasal cavity initially contacts the mucus layer lining the epithelium (Figure 3-1). Once in the mucus layer, formaldehyde undergoes a reversible reaction with water to form methanediol. Albumin in the mucus that lines the human nasal epithelium (Figure 3-1) forms an additional barrier to the systemic absorption of formaldehyde (Bogdanffy et al. 1987). The solubility of formaldehyde in mucus and the ciliary movement and ingestion of mucus may account for the removal of as much as 42% of the inhaled dose in rodents (Schlosser 1999). Diffusion is the dominant transport mechanism for formaldehyde through the mucus layer. Some inhaled formaldehyde passes through the mucus layer to reach the epithelium where its transformation and removal occur by enzymatic reactions with the nasal tissue and nonenzymatic reactions with glutathione and macromolecules, including proteins and DNA.

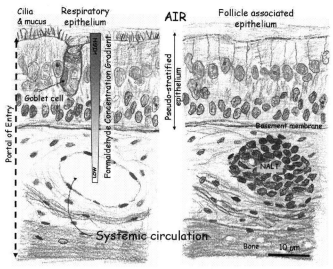

**FIGURE 3-1** Schematic representation of the mammalian nasal epithelium. Inhaled formaldehyde reacts rapidly with macromolecules in the tissue and the albumin in the mucus that lines the respiratory epithelium; these reactions result in a steep concentration gradient. Formaldehyde crossing the basement membrane can react further with macromolecules in the submucosal layer or reach the systemic circulation. The figure also shows a representation of the nasal-associated lymphoid tissue (NALT) that is generally present near the ethmoid turbinates on either side of the nasal septum and near the ventral nasopharyngeal duct. The NALT is one putative site of formaldehyde interactions with lymphoid tissues, but direct evidence that supports this hypothesis is lacking.

Removal of formaldehyde from the air by the upper respiratory tract is efficient. For example, nasal uptake of inhaled formaldehyde in dogs is nearly complete (Egle 1972). Penetration of formaldehyde to more distal airways was not observed in formaldehyde-exposed dogs even in the presence of increased breathing rates or exposure to formaldehyde at high concentrations (Egle 1972). In rats and mice, which are obligate nasal breathers with a highly complex nasal-airway geometry and large ratio of surface area to lumen volume, most inhaled formaldehyde (average, about 97%) is also taken up in the nasal cavity (Patterson et al. 1986). Computer models predict that systemic delivery of formaldehyde in rodents is extremely low and would not increase formaldehyde concentrations in tissues (Franks 2005). The results of those simulations are supported by animal studies that show that formaldehyde inhalation in rodents or nonhuman primates did not alter blood formaldehyde concentrations. Studies using DPX formation as an exposure surrogate corroborate that conclusion. For example, Casanova-Schmitz et al. (1984) found that exposure of rats to $^{14}$C- and $^{3}$H-labeled formaldehyde at 15 ppm did not result in DPX formation in bone marrow. However, they did observe non-linear DPX formation in the nasal mucosa at concentrations of 2 ppm or greater.

In humans, oral breathing bypasses the uptake in the nasal epithelium; this breathing pattern leads to increased delivery of formaldehyde to the intermediate regions of the respiratory tract (that is, from the oral cavity to the upper conducting airways of the lungs) (Overton et al. 2001). Nasal uptake of formaldehyde is also reduced as ventilation rates increase.

The committee concludes that the immediate fate of inhaled formaldehyde is primarily absorption at the site of first contact where it undergoes extensive local metabolism and reactions with macromolecules, despite species differences in uptake resulting from different breathing patterns (for example, oronasal breathing in humans vs nasal breathing in rodents). Although the site of absorption in the respiratory tract may depend on airway anatomy (simple vs complex), breathing pattern (nasal vs oronasal), and ventilation rate, the net result is that inhaled formaldehyde predominantly remains in the respiratory epithelium.

**Is Inhaled Formaldehyde Available Systemically?**

As mentioned earlier, several studies have demonstrated that inhaled formaldehyde has at most little systemic bioavailability. Some experiments examined the disposition of $^{14}$C-labeled formaldehyde, and others used techniques of analytic chemistry—such as gas or liquid chromatography coupled to mass spectrometry (for example, GC-MS)—to evaluate changes in formaldehyde concentrations in blood or tissue after inhalation. The studies have found that formaldehyde undergoes rapid elimination in blood with virtually no increase in "free" formaldehyde in blood or systemic tissues. Although the studies demonstrate relatively long retention of the $^{14}C$ radiolabel, this observation most likely reflects incorporation and turn over of the so-called 1-carbon pool. As noted by EPA, "measuring the distribution of the absorbed formaldehyde based on $^{14}C$-radiolabeling and GC-MS studies alone is problematic because it is difficult to resolve (through these studies) whether it is free, reversibly bound, irreversibly bound, formate, one-carbon pool, etc. This is of significance with regard to understanding the availability of the absorbed formaldehyde" (EPA 2010, p. 3-12). The committee shares EPA's concern.

A series of studies using dual-labeled ($^{14}C/^{3}H$) formaldehyde in rats has been performed to address the analytic concern (Casanova-Schmitz and Heck 1983; Casanova-Schmitz et al. 1984). The draft IRIS assessment accurately summarizes the main conclusions reached from those experiments, namely that "labeling in the nasal mucosa was due to both covalent binding and metabolic incorporation," that "DPX [were] formed at 2 ppm or greater in the respiratory mucosa," and that "formaldehyde did not bind covalently to bone marrow macromolecules at any exposure concentration" (up to 15 ppm) (EPA 2010, p. 3-12). The labeling of bone marrow macromolecules was found by the investigators to be due entirely to metabolic incorporation of the radiolabels, not to direct covalent binding of intact formaldehyde. The committee views those findings as supporting the hypothesis that inhaled formaldehyde is not delivered systemically under the exposure conditions used in the studies (0.3-15.0 ppm, 6 hr) (EPA 2010).

The committee also found that the more contemporary work performed by Lu et al. (2010) that examined formaldehyde-induced DNA adducts and DDX cross-links provided no direct evidence of systemic availability of inhaled formaldehyde. The Lu et al. (2010) study used $^{13}CD_2$-labeled formaldehyde and showed that $^{13}CD_2$-formaldehyde-DNA adducts and DDX were confined to the nasal cavity of exposed F344 rats, even though they examined much more DNA isolated from bone marrow, lymphocytes, and other tissues at distant sites for the adducts. The male Fischer 344 rats were exposed to [$^{13}CD_2$]-formaldehyde at 10 ppm for 1 or 5 days (6 hr/day) with a single nose-only unit.

The strongest data cited by EPA in support of systemic delivery of inhaled formaldehyde come from several studies in which antibodies to formaldehyde-hemoglobin and formaldehyde-albumin adducts were detected in blood from exposed workers, smokers, and laboratory animals. The studies did not definitively demonstrate, however, whether adduct formation occurs at a site distant from the portal of entry. For example, it is not known whether the adducts could be formed in the airway submucosal capillary beds or reflect systemic delivery of formaldehyde. Moreover, the draft IRIS assessment does not evaluate the antibody work as critically as the direct chemical-analysis approaches. The committee found that the draft does not offer a sufficient basis for EPA's reliance on the antibody data to support the hypothesis that formaldehyde (or its hydrated form, methanediol) may reach sites distal to the portal of entry and produce effects at those sites.

Questions have arisen regarding the possibility that formaldehyde reaches distal sites as methanediol. However, although equilibrium dynamics indicate that methanediol would constitute more than 99.9% of the total free and hydrated formaldehyde, the experimental data described above provide compelling evidence that hydration of formaldehyde to methanediol does not enhance delivery of formaldehyde beyond the portal of entry to distal tissues. Furthermore, Georgieva et al. (2003) used a pharmacokinetic modeling approach that explicitly accounted for the competing processes of hydration, dehydration, diffusion, reactivity with macromolecules, and metabolism and demonstrated that hydration-dehydration reaction rates determined from equilibrium studies in water are not applicable in biologic tissues, given that their use in the model resulted in simulations that were inconsistent with the available data. For example, the calculated dehydration rate from equilibrium dynamics studies in water was so small relative to other competing rates that too little formaldehyde would be available to account for the measured DPX rates. Thus, the data provide a strong indication that the hydration-dehydration reaction should not be rate-limiting and can thus be ignored in modeling the disposition of inhaled formaldehyde in nasal tissues.

EPA also suggested that systemic delivery of formaldehyde-glutathione adducts and latter release of free formaldehyde may result in delivery of formaldehyde to sites distal to the respiratory tract. However, experimental data supporting that hypothesis are lacking, as acknowledged by the draft IRIS assessment. In fact, additional data based on even more sensitive analytic methods published since the draft assessment was released casts further doubt on the hypothesis that formaldehyde reaches the systemic distribution in a form that can react with macromolecules in tissues remote from the portal of entry (Lu et al. 2011; Moeller et al. 2011; Swenberg et al. 2011).

The committee also found two divergent statements regarding systemic delivery of formaldehyde in the draft IRIS assessment. Some parts of the draft assume that the high reactivity and extensive nasal absorption of formaldehyde restrict the systemic delivery of inhaled formaldehyde to the upper respiratory tract (for example, EPA 2010, p. 4-371). Under that assumption, systemic responses—including neurotoxicity, reproductive toxicity, and leukemia—are unlikely to arise from the direct delivery of formaldehyde (or methanediol) to a distant site in the body, such as the brain, the reproductive tract, and the bone marrow. Other portions of the document presume systemic delivery of formaldehyde (or its conjugates) and use this presumption to account in part for the systemic effects (see, for example, p. 4-1, lines 16-19; p. 4-472, line 18; Section 4.5.3.1.8; and p. 6-23, line 31). The committee found the inconsistency to be troubling, and the divergent assumptions are not justified.

The committee concludes that the issue of whether inhaled formaldehyde can reach the systemic circulation is extremely important in assessing any risk of adverse outcomes at nonrespiratory sites associated with inhalation of formaldehyde. Moreover, the committee concludes that the weight of evidence suggests that it is unlikely for formaldehyde to appear in the blood as an intact molecule, except perhaps after exposures at doses that are high enough to overwhelm the metabolic capability of the tissue at the site of entry. Thus, although many sensitive and selective investigative approaches have been used, systemic concentrations from inhaled formaldehyde are indistinguishable from endogenous background concentrations. The committee, however, notes the importance of differentiating between systemic delivery of formaldehyde and systemic effects. The possibility remains that systemic delivery of formaldehyde is not a prerequisite for some of the reported systemic effects seen after formaldehyde exposure. Those effects may result from indirect modes of action associated with local effects, especially irritation, inflammation, and stress.

## Can Formaldehyde-Related Effects Alter Its Toxicokinetics?

Formaldehyde is an irritant gas in humans and animals. Rodents exposed to sufficiently high formaldehyde concentrations develop reflex bradypnea, decreased body temperature, reduced oxygen demand, and reduced nasal mucociliary clearance. For example, the minute volume decreased by 45% in rats and 75% in mice during formaldehyde inhalation at 15 ppm for 10 min compared with controls (Chang et al. 1983). The species difference may contribute to the difference between rats and mice in the incidence of nasal lesions. At identical exposures, mice were found to receive a lower effective dose at the target tissue in the nasal cavities than rats because mice have a greater reduction in minute ventilation in response to sensory irritation of the respiratory tract. The lower effective dose of mice was verified by the observation that mice had smaller increases in formaldehyde-induced cell proliferation in the nasal mucosa than rats. The findings could explain why higher formaldehyde exposure concentrations were needed to induce the same degree of toxic effects (such as nasal tumors) in mice as occurs in rats at lower exposures (Chang et al. 1983). The committee notes that the CFD models of Kimbell et al. (2001a,b) do not account for potential effects of sensory irritation on ventilation inasmuch as only two mass-transfer coefficients, one for mucus-coated and one for non-mucus-coated epithelial regions of the nose, were used in all simulations to derive uptake into nasal tissues. However, later models that account for DPX cross-links and cytotoxicity (Conolly et al. 2000, 2002, 2003, 2004; Georgieva et al. 2003) relied on animal data that were obtained at concentrations that potentially caused irritation to derive parameters associated with metabolism and reactivity; thus, the potential effect of altered ventilation was indirectly compensated for in those model simulations.

EPA also hypothesized that airway remodeling induced by formaldehyde exposure might alter formaldehyde dosimetry. For example, squamous epithelium absorbs considerably less formaldehyde than other epithelial types (Kimbell et al. 1997). Metaplasia of the anterior nasal epithelium to a squamous epithelial phenotype occurs in rats exposed repeatedly to formaldehyde at 3 ppm or higher (Kimbell et al. 1997). Although EPA identified the consequences of metaplasia for dosimetry, the committee concludes that this issue is not particularly important at the low exposure concentrations relevant to derivation of a reference concentration (RfC).

The committee agrees with EPA's conclusion that "certain formaldehyde-related effects have the potential to modulate its uptake and clearance" (EPA 2010, p. 3-5). Some of the effects, such as changes in mucociliary function and altered nasal epithelium, could occur in humans. However, reflex bradypnea and related modulating effects seen in rodents do not occur in phylogenetically higher animals (nonhuman primates) or humans. Thus, formaldehyde exposures at concentrations relevant for an RfC or unit risk are unlikely to alter its toxicokinetics.

**Are Formaldehyde-Induced Cross-Links Useful Biomarkers of Exposure?**

As noted earlier, formaldehyde readily forms DPX and DDX cross-links. A variety of analytic chemistry approaches, including radiolabeled formaldehyde tracers, have been used to evaluate formaldehyde binding to macromolecules and to differentiate binding from metabolic incorporation. For example, rats given intravenous injections of $^{14}$C-formaldehyde or $^{14}$C-formate, a metabolite of formaldehyde that contributes to the one-carbon pool, develop similar blood radioactivity profiles; this suggests that labeling of blood macromolecules is due to metabolic incorporation rather than adduct formation (Heck and Casanova 2004).

DPX formation in human white blood cells after in vitro exposure to formaldehyde (0-10 mM) for 1.5 hr rose linearly with increasing formaldehyde concentrations above 0.001 mM (Shaham et al. 1996). The kinetics of DPX formation in vivo (and in some in vitro systems) appear more complicated and reflect a balance between DPX formation and repair processes. Overall, DPX have been detected in upper and lower respiratory tracts of rodents and nonhuman primates. For example, Casanova and colleagues (Casanova et al. 1989, 1994) conducted formaldehyde inhalation studies in male F344 rats to determine DPX formation. In one study, DPX concentrations were measured from the nasal lateral meatus, medial meatus, and posterior meatus after inhalation of $^{14}$C-formaldehyde (Casanova et al. 1994). The sites were selected because they were associated with a high tumor incidence in a formaldehyde bioassay (Monticello et al. 1989, 1991). The other study measured DPX concentrations in nasal mucosal tissue taken from the entire nasal cavity (Casanova et al. 1989). Formation of DPX demonstrated a nonlinear dose-response relationship within the nasal epithelium; statistically significant increases occurred after exposure at concentrations of 0.3 ppm and higher.

Although the mechanisms for DPX-induced cell death have not been established, DPX formation has been used as a key event linking external exposures with cellular responses (mutagenicity, cytotoxicity, and compensatory cellular regeneration) in the development of pharmacokinetic and BBDR models. Pharmacokinetic models that describe in vivo DPX formation have included saturable pathways to describe enzymatic formaldehyde metabolism, a first-order pathway to represent formaldehyde's reaction with tissue constituents, and first-order binding to DNA (Hubal et al. 1997; Conolly et al. 2000). Yang and co-workers (2010) have suggested that a single, saturable pathway competing with DNA binding could describe the DPX data fairly well in rodents exposed to formaldehyde at 15 ppm or less. The models predict that less than 1% is covalently bound as DPX in the airway (Casanova et al. 1991; Heck and Casanova 1994).

DPX have also been reported in circulating lymphocytes from formaldehyde-exposed people (Shaham et al. 1996, 1997, 2003). For example, Shaham et al. (1996) measured DPX concentrations in human white blood cells from 12 workers exposed to formaldehyde and from eight controls. The authors found a statistically significant difference in the DPX concentrations in lymphocytes between the formaldehyde workers and the unexposed controls. However, only four of the 12 exposed workers had DPX concentrations above the upper range of the controls. The authors also reported a linear relationship between years of exposure and the amount of DPX and concluded that DPX may be used as a biomarker of formaldehyde exposure. Several limitations of the study have been identified by others (Casanova et al. 1996). Most notably, the control group's exposure history, including smoking prevalence, was poorly defined. Shaham et al. (1996) presented minimal exposure-assessment data, so the study was not used quantitatively by EPA. The committee agrees with EPA's decision.

The concentrations of DPX formed by formaldehyde in nasal airways have been modeled and used by EPA as an internal dose surrogate to update its health assessment of formaldehyde (EPA 1991; Hernandez et al. 1994). On the basis of different model assumptions that are discussed in more detail later in this chapter, EPA changed the values of the parameters in the DPX model (Subramaniam et al. 2007), which was used to derive internal dose-related points of departure for human extrapolation. On the basis of the evidence, the draft IRIS assessment states that "DNA protein cross-links (DPXs) formed by formaldehyde (covalently bound in this case) have been regarded as a surrogate dose metric for the intracellular concentration of formaldehyde [Casanova et al. 1989, 1991; Hernandez et al. 1994]. This is particularly relevant because of the nonlinear dose response for DPX formation due to saturation of enzymatic defenses at high concentrations [Casanova et al. 1989, 1991]. Thus, the ability to measure DPX is an important development" (EPA 2010, p. 3-12). The committee agrees with the use of DPX as a biomarker of exposure in the draft IRIS assessment.

**Can Inhaled Formaldehyde Have Systemic Genotoxic Effects?**

Formaldehyde is a genotoxic (DNA-reactive) chemical. Formaldehyde-induced DNA damage is postulated to lead to mutations and clastogenesis, critical cytogenetic events in the carcinogenic mode of action. The evidence of formaldehyde genotoxicity includes DPX cross-links, chromosomal aberrations, micronuclei, and sister-chromatid exchanges. A large number of in vitro tests for genotoxicity—including bacterial mutation, DNA strand breaks, chromosomal aberrations, and sister-chromatid exchange assays—are positive when formaldehyde is used. The studies on genotoxicity of formaldehyde, particularly those involving in vivo exposures of humans and animals, have provided strong evidence that formaldehyde genotoxicity occurs in the nasal mucosa and peripheral (circulating) blood lymphocytes. The database on genotoxicity assessment using peripheral blood lymphocytes from exposed human cohorts is appreciable and contains studies from different countries and various exposure scenarios and spans a period of more than 20 years. Some more recent studies (Costa et al. 2008) provide evidence of clastogenicity of formaldehyde coupled with individual exposure assessment. Although studies in humans showed some inconsistent results regarding the extent and form of the cytogenetic changes associated with formaldehyde exposure, the overall body of evidence suggests that inhaled formaldehyde has an effect that may be detected in blood cells in the systemic circulation. The committee notes that it is unknown whether formaldehyde genotoxicity arises from interactions that occur at the site of contact—for example, in nasal-associated lymphoid tissue in the nasal mucosa (Figure 3-1)—or as the result of local circulation of lymphocytes in blood that perfuses portal-of-entry tissues.

Overall, the draft IRIS assessment concludes that formaldehyde may act through a mutagenic mode of action and that this effect is not restricted to the site of entry (EPA 2010). As noted earlier, multiple lines of evidence support the conclusion that formaldehyde is mutagenic. The committee agrees with EPA and further notes that EPA's conclusion is consistent with the current *Guidelines for Carcinogen Risk Assessment* (EPA 2005).

The committee acknowledges that the database on the cytogenetic effects of formaldehyde in humans is supportive of EPA's second conclusion, that the mutagenic action of formaldehyde is not restricted to tissues at the point of contact. However, available data are insufficient to support definitive conclusions on several key issues. First, exposure assessment in the relevant human studies was generally lacking, and the effects observed occurred in highly exposed workers. In the absence of understanding of the shape of the dose-response curve for cytogenetic changes at low doses, it is difficult to extrapolate the findings to environmental exposures. Second, the mechanism of cytogenetic effects in circulating blood cells is not established—an uncertainty that complicates the committee's ability to link exposure with effects at distant sites. That data gap is especially problematic given the growing body of evidence that formaldehyde is not available systemically in any reactive form. Thus, it can only be hypothesized that systemic effects, such as cytogenetic effects in circulating blood lymphocytes, originate by as yet unproven mechanisms in portal-of-entry tissues.

The committee concludes that there is great uncertainty and associated controversy regarding the following issues: the ability of formaldehyde to cause DNA damage at distal (that is, other than portal-of-entry) sites, the relative contributions of DNA adducts vs cross-links to carcinogenesis, the ability of formaldehyde exposures at ambient concentrations to increase the burden of endogenous formaldehyde-induced DNA damage that can contribute to carcinogenesis at local or distant sites, and the consistency of the database on formaldehyde-induced genotoxicity among species.

## Are Useful Computational Pharmacokinetic Models for Formaldehyde Inhalation Available?

EPA's *Guidelines for Carcinogen Risk Assessment* state that "toxicokinetic modeling is the preferred approach for estimating dose metrics from exposure. Toxicokinetic models generally describe the relationship between exposure and measures of internal dose over time. More complex models can reflect sources of intrinsic variation, such as polymorphisms in metabolism and clearance rates. When a robust model is not available, or when the purpose of the assessment does not warrant developing a model, simpler approaches may be used" (EPA 2005, p. 3-1). Anatomically based three-dimensional (3D) computational-fluid-dynamics (CFD) models of rat, monkey, and human nasal passages have been developed to predict interspecies nasal dosimetry of inhaled formaldehyde (Table 3-1). Mass-transfer coefficients calibrated against total nasal uptake were used as boundary conditions in CFD models to determine site-specific formaldehyde flux rates (Kimbell et al. 1993, 2001a,b). As is consistent with experimental data, regional flux rates predicted by the models depend on airflow characteristics, exposure concentration, and the absorption properties of the nasal lining.

CFD model estimates of nasal uptake can range from about 60% to more than 90% of inhaled formaldehyde depending on exposure conditions and species. Reduced nasal uptake does occur at high air concentrations, but this would not be predicted with the current CFD model (Kimbell et al. 1993, 2001a,b), because it did not include a saturable mechanism for tissue metabolism of formaldehyde in the mass-transfer boundary condition. In humans, about 90% absorption of inhaled formaldehyde is predicted to occur in the nose on the basis of a single CFD model with pharmacokinetic parameters scaled from animals at resting ventilation; this estimate decreases to about 60% with light exercise and 55% with heavy exercise (Kimbell et al. 2001b).

CFD models were combined with biologically based pharmacokinetic (PK) models to describe site-specific and regional uptake of formaldehyde along conducting airways of rats and monkeys at sites that correspond to areas that have measured DPX, cell proliferation, pathology, or tumor formation (Casanova et al. 1991; Kimbell et al. 1993; Hubal et al. 1997; Conolly et al. 2000; Kimbell et al. 2001a,b). Some models incorporate the flux of formaldehyde into cells of nasal passages as the model input (Hubal et al. 1997; Conolly et al. 2000). The most advanced models provide an estimate of cellular formaldehyde

**TABLE 3-1** Analysis of 3D CFD Models by Kimbell et al. (2001a,b) and Overton et al. (2001) for Rat, Monkey, and Human Airways

| Component | Model Strengths | Model Weaknesses |
|---|---|---|
| CFD model design | • Incorporated airway anatomy used to define species and site-specific airflows in the nose<br>• Calibrated with experimental data based on molds and water flows | • One geometry used per species; individual variability not accounted for<br>• Relatively poor resolution in serial histologic data, which leads to unrealistic, jagged-surface meshes that contribute to mass-balance errors<br>• Only one side of nasal airways of the rat and monkey are used (symmetry assumed)<br>• Model did not include external nares, flexible nasal walls, mucus movement, nasal hairs, or water vapor<br>• The mucus layer was assumed to be uniform over all surfaces except the vestibule<br>• Uncertainties in the estimation of surfaces associated with mucus-coated vs non-mucus-coated airways |
| Site-specific fluxes of formaldehyde | • Compared well with the distribution of lesions in toxicity studies and cell-proliferation data<br>• Only two mass-transfer coefficients (one for mucus-coated and one for non-mucus-coated nasal regions) were used; thus, site specificity to fluxes in these two types of regions were driven only by local airflows<br>• Mass-transfer coefficients were based on a single, assumed value of mucus thickness (20 μm); any changes to these calculations would result in altered flux determinations | • Only steady-state inhalation simulations conducted rather than transient, full breathing cycle<br>• The element-by-element contribution to formaldehyde flux errors was greatest in regions of more complex geometry and lower in regions with minimal topographic changes<br>• A study by Bogdanffy et al. (1986) suggests that the distribution of aldehyde dehydrogenase activity is not uniform in all regions of mucus-coated epithelium; this could affect regional flux rates<br>• No data exist for comparing site-specific fluxes of formaldehyde with model results (model resolution is substantially greater than experimental resolution); model evaluations were based on secondary or downstream biomarkers of dose, such as cell proliferation, and lesion-mapping flux rates on the surface can be sensitive to metabolism, reactivity, blood flows, and other mechanisms of tissue clearance |

*(Continued)*

**TABLE 3-1** Continued

| Component | Model Strengths | Model Weaknesses |
|---|---|---|
| | | • Clearance is influenced by estimates of tissue or mucus thickness at higher concentrations |
| | | • The computational mesh used for human simulations did not allow reasonable convergence of the flow equations solved at the highest flow rates for heavy exercise |
| | | • Relationships between flux rates and cytotoxicity and DPX formation are clearly nonlinear; thus, local variation or error in flux predictions can have substantial effects on predictions of cytotoxicity and DPX in BBDR models |
| Human respiratory tract | • CFD-derived fluxes of formaldehyde in nasal airways of humans under several activity patterns (respiratory rates) are used to calibrate the nasal compartments of a single-path, one-dimensional anatomic model of the human respiratory system that is then used to calculate fluxes in all airways<br>• Airways beyond the rigid trachea expand and contract to greater degrees (increase in compliance) as air (tidal volume) moves distally, allowing the model to simulate the full breathing cycle | • Idealized model is based on the recursive, symmetrical bifurcating lung tree of Weibel and not on real airway geometry<br>• Single mass-transfer coefficient for formaldehyde was assumed to be the same in all airways |
| CFD-PK models | • Incorporate actual histologic measurements of distances between the air interface and DNA (nuclei), the basal membrane, and bone from control rats to develop the model for respiratory and transitional epithelium | • Model tracks only total formaldehyde with the understanding that this represents mostly methanediol, and it uses the calculated diffusivity constant for methanediol rather than free formaldehyde in simulations of nasal mucosa<br>• Metabolism of formaldehyde by formaldehyde dehydrogenase (now known as ADH3) and other aldehyde dehydrogenase enzymes was described with a single empirical term because no data existed to differentiate the rates in vivo |

Abbreviations: CFD, computational fluid dynamics; PK, pharmacokinetic; DPX, DNA-protein crosslinks; BBDR, biologically based dose-response; and ADH3, alcohol dehydrogenases.

concentrations based on CFD model predictions of flux, estimated epithelial surface areas, assumed tissue thicknesses in low-tumor and high-tumor regions, and the rate of formaldehyde reaction with cellular constituents via a saturable enzyme-mediated metabolism, a nonenzymatic first-order reaction with cellular macromolecules, or a first-order reaction with DNA to form DPX. Once formed, DPX were eliminated actively or by normal degradation at a constant rate. The models also predict a steep, decreasing concentration gradient between the mucus layer and each successive layer of cells from the airway to the vasculature as a result of formaldehyde reaction with extracellular and cellular glutathione, proteins, other macromolecules, and metabolizing enzymes. The draft IRIS assessment provides a comprehensive overview of the CFD PK models that are available for formaldehyde.

The draft IRIS assessment raises the criticism that the nasal CFD models are based on a single geometry for each species. Thus, the models do not address variability that arises from differences in airway anatomy. A recent paper by Garcia et al. (2009) evaluated the effect of individual differences in airway geometry on airflow and uptake of reactive gases, such as formaldehyde. Although the sample was small (five adults and two children), the individual differences in airway geometry alone caused the potential flux rates to vary by a factor of only 1.6 over the entire nose and by a factor of 3-5 at various distances along the septal axis of the nose. The committee agrees with EPA that although the sample was small, the estimates of individual variability are consistent with default uncertainty factors applied to internal dose metrics that account for human variability. Another EPA criticism is that idealized geometries (cylinders) rather than real geometries are used to represent the human larynx, trachea, and lung with no evaluation of potential variability associated with actual airway geometry.

EPA also raised a concern that high formaldehyde concentrations (3 ppm or higher) can reduce minute volumes, alter mucus flow, or change absorption by tissue remodeling and that the existing models capture these effects inconsistently. The committee notes that mass-transfer coefficients in the CFD models are used to capture the effects of exposure at irritating concentrations (15 ppm) at which respiratory depression occurs (Kimbell et al. 2001a,b); this allows the models to describe the experimental data adequately at high exposure concentrations. However, the committee also notes that the mass-transfer coefficients used in the human models were optimized on the basis of the rat models.

Despite the concerns raised, EPA used the CFD models to derive human-equivalent concentrations (HECs) on the basis of formaldehyde flux rates. The CFD-based flux rates over the entire nose of the rat (excluding the vestibule and olfactory region) were used to calculate the HEC by multiplying the average flux for the rat at a no-observed-adverse-effect level by a dose-duration adjustment—(5/7)(6/24)—to represent continuous exposures. EPA also calculated benchmark concentrations by using flux rate as the internal dose in dose-response assessments. The models were used to evaluate whether the lower airways in the human are potential targets for formaldehyde toxicity and carcinogenicity. To do that, EPA used the one-dimensional human-airway model of Overton et al. (2001) to determine the risk of lower respiratory airway tumors, assuming equal sensitivity of cells in this region to equivalent formaldehyde fluxes determined for the nose. EPA restricted application of the models to the experimental range used in animal studies to determine an internal dose-based point of departure for humans but did not use them to extrapolate to low exposures.

The committee disagrees with EPA's findings that CFD models are not useful for low-dose extrapolations. In fact, flux results from the CFD models can easily be scaled from an exposure of 1 ppm—as given by Kimbell et al. (2001a,b) and Overton et al. (2001)—to lower concentrations because of the linear flux-concentration relationship that was used by the authors. Therefore, the committee recommends that the CFD-based approach also be used to extrapolate to low concentrations, that the results be included in the overall evaluation, and that EPA explain clearly its use of CFD modeling approaches.

The committee concludes that sufficiently robust pharmacokinetic models for formaldehyde exist and agrees with EPA that the CFD models can and should be used in the IRIS assessment. Furthermore, it finds that the CFD models were fairly evaluated and that the sources of uncertainty in dose metrics used in dose-response assessments were appropriately treated.

## CARCINOGENESIS: HAS A MODE OF ACTION OF FORMALDEHYDE BEEN IDENTIFIED?

*Mode of action* is defined by EPA as a sequence of key events and processes, starting with the interaction of an agent with a cell and continuing through operational and anatomic changes that result in an adverse outcome (EPA 2005). A key event is an empirically observable precursor step that is itself a necessary element of the mode of action or is a biologically based marker of such an element (Boobis et al. 2009). Knowledge of the mode of action can inform the risk-assessment process. Indeed, mode of action in the assessment of potential carcinogens is a main focus of EPA's cancer guidelines (EPA 2005). In the absence of sufficient, scientifically justifiable information on mode of action, EPA generally takes public-health-protective, default positions regarding the interpretation of toxicologic and epidemiologic data.

The draft IRIS assessment provides an exhaustive summary of the studies on genotoxicity of formaldehyde. The literature review appears to be up to date and includes all major and recent studies. The relevant chapter is well organized by type of DNA damage and then by evidence of clastogenicity from in vitro and in vivo sources. More weight is placed on studies in human cells and in exposed human cohorts, however small each study may be. Data are presented in informative and well-organized tables that provide a high-level summary of the extensive database on each subject. The summary statement and the entire chapter are well balanced and include both positive and negative studies. The conclusion that formaldehyde is genotoxic and mutagenic in model systems and in mammals, including humans, is supported by the data and is in accordance with the weight of evidence required by EPA's cancer guidelines (EPA 2005).

Cytotoxicity and compensatory cell proliferation also appear to play important roles in the carcinogenic mode of action of formaldehyde-induced nasal tumors. Substantial nonlinearity in dose-response relationships was observed in animal studies of DPX cross-links, cytotoxicity, and compensatory cell proliferation (Swenberg et al. 1983; Monticello et al. 1996). There is a strong site concordance between formaldehyde uptake, cytotoxicity, cell proliferation, and tumor formation. Furthermore, no tumors were observed at concentrations that did not also cause cytotoxicity. The draft IRIS assessment discusses this alternative mode of action but relies on the mutagenic mode of action to justify low-dose linear extrapolations in the assessment of formaldehyde-induced nasal tumors.

In the case of hematopoietic cancers, particularly leukemia, much less is known about potential modes of action other than mutagenicity, which has been demonstrated in vitro and in a few studies of occupationally exposed humans. Although EPA postulated that formaldehyde could reach the bone marrow either as methanediol or as a byproduct of nonenzymatic reactions with glutathione, numerous studies described above have demonstrated that systemic delivery of formaldehyde is highly unlikely at concentrations below those which overwhelm metabolism according to sensitive and selective analytic methods that can differentiate endogenous from exogenous exposures. As a result, EPA could only speculate that circulating hematopoietic stem cells that percolate through nasal capillary beds or nasal-associated lymphoid tissues may be the target cells for mutations and clastogenic effects that eventually result in lymphohematopoietic cancers. Experimental evidence of either mechanism is lacking.

Although EPA followed its guidelines for assessing the risk of cancer associated with a mutagenic mode of action, it acknowledged that major uncertainties and controversy remain regarding application of linear models for low-dose extrapolations for a chemical that is formed endogenously and is too reactive to be measured in the body apart from portal-of-entry tissues. As discussed in the following section on BBDR modeling, the committee recommends that, for transparency and completeness, EPA consider providing alternative calculations that factor in nonlinearity associated with the cytotoxicity-compensatory cell-proliferation mode of action and assess the strengths and weaknesses of each approach.

The committee concludes that two primary modes of action have been observed to contribute to formaldehyde-induced carcinogenicity in nasal tissues: mutagenicity and cytotoxicity with compensatory cell proliferation. There is no doubt that formaldehyde is a DNA-reactive chemical that produces DNA adducts (DPX cross-links and DDX cross-links) that, if not repaired, can lead to mutations and clastogenesis.

## USE OF A BIOLOGICALLY BASED DOSE-RESPONSE MODEL

EPA's *Guidelines for Carcinogen Risk Assessment* state that "in the absence of sufficient data or understanding to develop...a robust, biologically based model, an appropriate policy choice is to have a single preferred curve-fitting model for each type of data set" (EPA 2005, p. 1-10). The guidelines acknowledge that many curve-fitting models have been developed and that the ones that fit the observed data reasonably well may lead to wide differences in estimated risk at the lower end of the observed range. The guidelines also recognize that several competing models could be developed and further state that "if critical analysis of agent-specific information is consistent with one or more biologically based models as well as with the default option, the alternative models and the default option are both carried through the assessment and characterized for the risk manager. In this case, the default model not only fits the data, but also serves as a benchmark for comparison with other analyses" (EPA 2005, p. 1-9). The committee notes that the use of default and alternative models for formaldehyde risk assessment remains controversial.

### What Is the Status of Biologically Based Dose Response Models for Formaldehyde?

A series of papers described the development of a biologically based model for formaldehyde in rats (Conolly et al. 2003) and humans (Conolly et al. 2004). A consistent focus of formaldehyde modeling has been the low-dose linear formation of DPX as a key component in the mode of action of cytotoxicity and carcinogenesis (Conolly et al. 2000). The model improves on previous CFD-derived estimates of nasal-airway flux and DPX formation in the anterior portion of the rat nose (Hubal et al. 1997) by using a more complete version of the rat-nose and nasal-airway models for the monkey and human as described by Kimbell et al. (2001a,b) and Overton et al. (2001). The whole-nose DPX data were supplemented with regional-DPX data on the F344 rat (Casanova et al. 1994) and rhesus monkey (Casanova et al. 1991) to correspond with areas of high and low tumor incidence. The key linkage (internal dose metric) between exposure and DPX is the concentration of formaldehyde in nasal tissue. The local tissue concentration is determined from CFD-derived formaldehyde flux rates, the thickness of the epithelium, and saturable metabolism and first-order clearance from the tissue. DPX concentrations are calculated by assuming first-order rates of binding to DNA and first-order rates of repair of DPX (thus, the low-dose linear relationship with exposure and higher than linear increase in DPX at concentrations above saturable metabolism by alcohol dehydrogenases). The Conolly et al. (2003, 2004) models can be used to predict the location and magnitude of cell proliferation in respiratory tissues in response to formaldehyde exposure conditions (concentration and duration) and respiratory rates (resting or working conditions).

Conolly and colleagues (2003) used rat data on both formaldehyde-induced DPX and cytotoxicity-compensatory cell proliferation as modes of action to link regional dosimetry with rat tumor responses. Dose-dependent tumor responses were predicted by using the Moolgavkar-Venzon-Knudsen (MVK) two-stage clonal-growth model of cancer (Conolly et al. 2003), in which DPX formation was assumed to increase the probability of mutations that lead to squamous-cell carcinoma in rat nasal tissue. The clonal-growth model includes mutation rates (mutagenic mode of action) of normal and intermediate cells and rates of birth and death of normal and intermediate cells (cytotoxicity-compensatory cell proliferation) that lead to the formation of malignant cells and, after a delay, tumor responses. The clonal-growth model was calibrated against data collected from a 2-year pathogenesis bioassay (Monticello et al. 1996) and extrapolated to humans to predict additional cancer risk associated with continuous environmental exposure to formaldehyde (Conolly et al. 2004).

The assumptions used by Conolly et al. (2003, 2004) to develop their models were extensively reviewed and evaluated by EPA for their potential effect on model predictions in the draft IRIS assessment (Table 3-2). For example, the only structural difference between the human model and the rat model is that the human model includes the one-dimensional full-respiratory-tract model of Overton et al. (2001) to determine flux rates as the dose metric. Rather than a focus only on predicting nasal tumors, the human model was used to predict the risk of all human respiratory tract tumors. The human-DPX model was scaled up from the rat and monkey models on the basis of allometric relationships of physiology and metabolism to body weight. For the human two-stage clonal-growth model, baseline mutation rates were calibrated against human lung-cancer incidence data, and formaldehyde-specific parameters associated with mutation rates and growth advantages of intermediate cells were assumed to be the same as those developed for the rat model. Although Conolly et al. (2003) clearly noted the need for additional research to address uncertainties and variability issues associated with this model (as is the case with any model), they believed that they made conservative choices—use of the hockey-stick model for cell proliferation rather than the best-fit J-shaped model and steady-state inhalation formaldehyde flux rates and oronasal breathing under various working conditions as the internal dose driver rather than realistic cyclic breathing rates.

A major strength of the human model is that its developers were consistently clear about the model structures, the assumptions used to simplify an otherwise complicated process, and how the model parameters were measured, fixed on the basis of relevant data or known system constraints, or estimated from in vivo data. The developers were also open about sources of variability and uncertainty in their models and further demonstrated the utility of a unifying model to help to identify aspects that have the most influence on predictions. Thus, the model is "transparent" (if technically challenging) and available for

**TABLE 3-2** Overview of the Conolly et al. BBDR Models

| Model Component | Model Assumptions | Potential Effect | Possible Model Refinements |
|---|---|---|---|
| Parameters associated with saturable metabolism, first-order clearance, and first-order DNA binding | Parameters were all optimized from the regional DPX concentration data in rats and monkeys rather than independently determined and then verified against the DPX data. Estimates of the first-order loss of DPX (repair) were arbitrarily set to the lowest value that ensured complete clearance of DPX in 18 hr for the highest exposure (that is, time from end of one 6-hr exposure at 15 ppm to start of next 6-hr exposure). | The first-order rate constant for DNA binding accounts for a small fraction of formaldehyde tissue clearance; thus, uncertainty is associated with estimates of saturable metabolism and the first-order clearance process that dominate the disposition of formaldehyde in respiratory tissues. | Additional experimental data would be needed to independently develop and evaluate the key process associated with formaldehyde disposition in nasal tissues. As such data are developed, fewer parameters would need to be optimized, and overall uncertainty in modeling tissue dose would be reduced. |
| Nasal blood flow | Blood flow in the nasal submucosa was not considered in the development of the BBDR models because existing data showed that no detectable increase in blood formaldehyde occurs after inhalation exposures. | The presence of albumin and hemoglobin adducts suggests that formaldehyde can penetrate into the blood perfusing the nasal submucosa (although formaldehyde's reactivity has precluded its penetration and detection in systemic blood). Thus, local blood flow could have an effect on the optimization of metabolism and clearance rate constants; this is a testable hypothesis. | Explicitly define regional nasal blood flow in the model and determine its sensitivity to current parameter estimates. Species-specific blood-flow data are sparse; a wide range of flows (0.1-1% of cardiac output) have been used in other models. |

| | | | |
|---|---|---|---|
| Tissue thickness | Use of measured tissue-thickness averages for the monkey gave visually poor fits to the DPX data during optimization of the clearance parameters. | Directed Monte Carlo approach was used to optimize tissue thicknesses to values within one standard deviation of the measured mean; this finding indicated that the parameters associated with the clearance and binding of formaldehyde are sensitive to the estimates of tissue thickness. | Better estimates of actual tissue thickness (and better dissection techniques for DPX measurements) would mean that this variable would not have to be optimized and would improve confidence in the other optimized parameters. |
| Flux bins | Nasal flux bins represent percentage of nasal surface areas that achieve a particular formaldehyde flux that fall within 20 equally divided intervals of flux ranging from 0 to the maximum rate on the airway surface; thus, no geometric location or site specificity is implied. | Flux bins are based on 20 equally spaced distributions of formaldehyde flux rates and not representative of specific locations in the nose. Thus, site-specific flux rates are not matched to site-specific DPX measurements to derive estimates of metabolism, binding, and clearance processes in the nose.<br><br>The 25 flux bins for the rest of the respiratory tract used in this model are associated with anatomic regions rather than predicted flux ranges. | Better definitions of CFD-model boundary conditions—including linkage to PBPK models for localized tissue concentrations in mucus, epithelium, and submucosa similar to that in models already developed for other reactive gases, such as acrolein and hydrogen sulfide—would improve the optimization of metabolism, binding, and clearance parameters and reduce the uncertainty by improving the correlation between regional dosimetry and DPX binding data. |
| Two statistical models were evaluated that relate time- and site-averaged cell proliferation data to formaldehyde exposure | J-shaped model provided the best fit to the data.<br><br>Hockey-stick model was considered more conservative because it removed the low-dose decrease in cell proliferation and | The J-shaped model implies that at low formaldehyde exposure concentrations, cell proliferation is less than control levels. Although that model provided a better description of the data, its application in the BBDR models would result in | None; both models (J-shaped and hockey-stick) should be carried through the simulations as was done by Connolly et al. The simulations should be compared with the EPA default no-threshold low-dose extrapolation assumption, and the results clearly |

*49*

*(Continued)*

**TABLE 3-2** Continued

| Model Component | Model Assumptions | Potential Effect | Possible Model Refinements |
|---|---|---|---|
| | replaced it with a threshold that was fixed at the lowest concentration that produced a measurable increase in cell proliferation. | fewer tumors predicted at low exposures than predicted in controls. Thus, the hockey-stick model, which implies a threshold for tumors, was also carried forward in the BBDR models. | compared and evaluated as to the relative strengths, weaknesses, and effects of each approach. |
| Formaldehyde flux rates derived from the CFD models used to provide input into the BBDR models for rats and humans | Same assumptions, strengths, and weaknesses of the CFD models apply to the Conolly et al. models (see Table 3-1) | Flux rates were based partly on two mass-transfer coefficients (one for mucus-coated and one for non-mucus-coated epithelium) that were estimated from data on nasal extraction of inhaled formaldehyde in rats. The mass-transfer coefficients were assumed to be constant across ventilation rate, formaldehyde concentration, region of the nose, and species. Because regional and species differences in metabolism would be expected, site-specific flux determinations used as inputs into the BBDR models would also be expected to be affected by the assumed mass-transfer coefficients. | See Table 3-1 and previous refinements in which direct coupling of PK models as boundary conditions on the CFD model would decrease the uncertainties associated with site-specific estimates of metabolism, binding, and clearance parameters that affect the inputs into the BBDR models. |

Abbreviations: DPX, DNA-protein crosslinks; BBDR, biologically based dose-response; CFD, computational fluid dynamics; PK, pharmacokinetic; PBPK, physiologically based pharmacokinetic; and EPA, Environmental Protection Agency.

evaluation and testing of alternative approaches (as was done by EPA). The committee examined key assumptions associated with the Conolly et al. models and estimated the likely effect of changes in the assumptions on model output (Table 3-3).

EPA agreed that the Conolly et al. (2003) model provided a better fit to the time-to-tumor data in the animal bioassays than other models that EPA could use in a cancer risk assessment. EPA also expressed confidence in the dosimetry modeling for flux rates and DPX formation in the rat and monkey, although it expressed concern that one model is used as representative of a population and another is an idealized model for the human lung.

EPA largely disagreed with many of the other conclusions presented in Conolly et al. (2003). It did not agree with the interpretation that DPX formation (and therefore mutagenicity) at low doses was negligible and did not contribute to an increase in nasal cancer risk. The committee notes that although Conolly and co-workers (2003, 2004) assumed that cytotoxicity-compensatory cell proliferation was the dominant mode of action in predicted tumor responses, they were careful to use an upper bound on values of rat parameters to force the model to calculate additional risk due to other mechanisms, such as mutagenicity.

EPA conducted a reanalysis of the Conolly et al. models, which was reported in a series of papers published in 2005-2010 (Crump et al. 2005, 2008, 2010; Subramaniam et al. 2007, 2008), and explored the uncertainties and sensitivities of the dynamic response components of the Conolly et al. formaldehyde models to changes in several key model parameters and assumptions. EPA's reanalysis was consistent with its cancer guidelines that specify that the uncertainties and variability in model parameters must be understood and articulated so that predictions of adverse responses and extrapolations to human exposures can be appropriately characterized from the standpoint of human health protection (EPA 2005). Thus, EPA's reanalysis focused on several key components in the Conolly et al. models that can have substantial effects on predicted human tumor formation and low-dose extrapolations when the models are used below the range of experimental observations.

EPA's analysis (EPA 2010) evaluated the following:

- Choice of background nasal-tumor incidence data in rats and total respiratory-tumor incidences in humans, which were used to define basal mutation rates for normal and intermediate cells.
- Choices of model structure and associated parameters used to describe DPX-formation data and cell-proliferation data collected in rats and monkeys as a function of time and exposure (that is, linear vs J-shaped vs hockey-stick-shaped dose-response curves).

**TABLE 3-3** Effects of Different Parameters on Predicted Results of the Conolly et al. BBDR Models

| Parameter Assumptions | Likely Effect on Model |
|---|---|
| Regenerative cell-proliferation data were directly related to cytotoxicity | Moderate; the model results were sensitive to the model parameters used, but the assumption that regenerative cell proliferation is directly related to formaldehyde-induced cytotoxicity is reasonable. |
| Site-to-site variation in cell proliferation in the rat nose does not vary in concordance with site-specific flux rates as determined by the CFD model | Moderate; only two mass-transfer coefficients were used to drive the uptake in all areas of the nose. That ignores the potential for site-specific local differences in tissue thickness and clearance processes that affect uptake. |
| Cell proliferation followed a J-shaped dose-response relationship | Minimal; although the cell-proliferation data suggest a J-shaped dose-response relationship, both a J-shaped and a hockey-stick model were used in the BBDR models. |
| Human cells were equally sensitive to the same internal dose surrogate as rats and were the basis of model parameters | Minimal, because the relationships between cell proliferation and simulated formaldehyde flux rates were similar in rats and monkeys, it is reasonable to assume that they would be similar in humans. Subramaniam et al. (2008) suggest that in the exposures at which tumors were seen, the mutagenic mode of action contributed up to 74% of the added tumor probability, whereas Conolly et al. (2003, 2004) showed little contribution by this mode of action in the human when they used an upper bound on the DPX proportionality constant rather than the maximum likelihood estimate of zero from the rat data. |
| Number of cells at risk was proportional to body weight | Unknown; it is not known how sensitive the BBDR models are to changes in the number of cells at risk by body weight. |
| DPX was a promutagenic lesion that was proportional to DPX burden | High; the current parameter estimates that Conolly et al. (2003) optimized from the data, using a maximum likelihood function, suggest that the proportionality constant for DPX adding to the mutation rate of a normal (or intermediate) cell should be zero or close to zero. That suggests that DPX is not directly related to the key events leading to mutation and carcinogenicity per se. Because this is the only low-dose linear relationship between exposure and a biomarker of response, EPA contends that the low-dose extrapolations should be linear through zero dose. For example, Subramaniam et al. (2007) examined alternative choices to parameters associated with DPX clearance and suggested that in the exposures at which tumors were seen, the mutagenic mode of action could contribute up to 74% of the added tumor probability. Because too few parameters were experimentally fixed and too many optimized against one data set, confidence in deciding whether the Conolly et al. or the Subramaniam et al. approach is the most scientifically defensible is not high. |

In a follow-on analysis, Crump et al. (2008) made an arbitrary change in the DPX-based effect on initiated cell replication by theorizing that if an initiated cell is created by a specific mutation that impairs cell-cycle control, there may be a mitigation of cell replication that is observed in the low-dose cell proliferation of normal cells (that is, in the negative vs baseline replication portion of the J-shaped dose-response curve) and hence a shift of the cell division of an initiated cell in the model toward greater rates at low doses. They argued that this theoretical possibility provides biologic motivation to test the effect on the MVK model because of the absence of data for doing otherwise. The change disconnects the birth and death rates of initiated cells from constraints used by Conolly et al. based on normal cells. The committee concludes that this change is contrary to the explanation provided by Monticello et al. (1996), who suggested that it is not a mutation in cell-cycle check points that results in lower cell-division rates than control at low exposures but rather an increase in the time that it takes for DNA-repair processes to eliminate the DPX before the cell can resume the process of cell division that leads to a cell-division rate at low exposures that is below that seen under basal conditions. These are two fundamentally different mechanisms with different connotations for risk—the mutagenic one chosen by EPA and the DNA-repair mode of action supported by several other publications on DPX cited by Conolly et al. (2003, 2004) and Monticello et al. (1996). Because they argue that there are no data to refute these assumed and arbitrary adjustments of the Conolly et al. models, they state that the onus is on others to show that such small changes cannot occur (that is, prove a negative before the authors would accept the contention that the Conolly et al. models are at all conservative as Conolly et al. suggested). That standard cannot be met.

High; the MVK model is very sensitive to the choice of controls and their effect on basal mutation rates.

There were zero squamous cell carcinomas in control rats in the two bioassays used to define the basal mutation rates of normal and intermediate cells in the two-stage, MVK dose-response model. Conolly et al. (2004) used results from the full National Toxicology Program historical control database. That is a point of contention by EPA, which believes that only historical controls from inhalation bioassays (and those in the same laboratory as the formaldehyde study) can be used in a relevant comparison. Squamous cell carcinomas are so rare that some leeway in approximating basal rates may have to be accepted, even though EPA's point is technically correct.

When mutation rates, cell-division rates, and mutation intensity vary as a function of age (that is, not treated as constants over all ages), the Hoogenveen solution can lead to more errors than alternative models that respect the nonhomogeneity of the model parameters (Crump et al. 2005). Although EPA may be technically

Basal mutation rates for normal and intermediate cells are unknown and had to be optimized from sparse information using several assumptions

*(Continued)*

TABLE 3-3 Continued

| Parameter Assumptions | Likely Effect on Model |
|---|---|
|  | correct, it was unclear to the committee whether there is a significant effect on the Conolly et al. (2004) prediction unless their parameter values are substantially changed, which would open the Crump et al. and Subramaniam et al. models up to an equally fair criticism of their assumptions and choices in model parameters. |
|  | Estimating parameters for basal mutation rates for a normal to intermediate and intermediate to malignant transformation in humans is subject to even more uncertainty than in the rat. |
|  | Crump et al. (2008) made seemingly arbitrary changes in the birth and death rates of initiated cells, claiming that there are no data to suggest that they cannot do that and that their changes were small relative to changes in this relationship at high formaldehyde flux rates. Although no changes in the predicted data range were observed, the effect of the low-dose and high-dose extrapolations was huge. |
| Intermediate-cell parameter values were related to normal-cell parameter values and constrained by relevant cell-proliferation and tumor data | Moderate. The proportionality constant for relating DPX to the human cell-mutation rate (KMU) was adjusted from the upper bounds on the rat value rather than the maximum likelihood estimate of the rat value, which was zero, by a ratio of the basal mutation rate of human to rat; this assumes that the mutation mechanism is the same in control vs formaldehyde exposure and that DPX can be viewed as a promutagenic event. Those assumptions suggest that human cells are more difficult to mutate than rat cells on the basis of their evaluation of the literature, and they force the model to include clonal growth at exposures below those known to cause cell proliferation in the rat. The authors view the assumptions as conservative. |
| Changes in epithelial cell thickness due to formaldehyde exposure had no effect on DPX measurements used to calibrate the BBDR models | Moderate. The first-order clearance of DPX could be slower than that used by Conolly et al. (2003, 2004). Over time, epithelial tissue in targeted regions of the nose thickens. The thickening could conceivably dilute DPX concentrations in the measured tissues to such an extent that residual concentrations 18 hr after exposure are not different from those in naïve animals, and this would affect the determination of DPX clearance rates. |

Abbreviations: DPX, DNA-protein crosslinks; BBDR, biologically based dose-response; CFD, computational fluid dynamics; EPA, Environmental Protection Agency; and MVK, Moolgavkar-Venzon-Knudson.

- Assumptions used in defining parameters associated with mutation, birth, and death rates of intermediate cells, cells that cannot be directly measured or even identified histologically (that is, intermediate cells in the MVK 2-stage model structure are a surrogate for potentially multiple stages of transformed but not yet malignant cells).

Other sources of variability and uncertainty were also explored, but the three noted were the main focus of EPA's reanalysis. The committee agrees that the sensitivity analysis added value to the interpretation of the Conolly et al. models, although it raised questions about the degree to which manipulations of the range of model parameter values can and should be performed to reflect potentially divergent outcomes. EPA's reanalysis also identified a flaw in one of the numeric approaches used in the original models and corrected it to improve the reliability of the simulation—another value added.

Alternative model structures fitted to the data used by Conolly et al. can yield different low-dose extrapolation results (Crump et al. 2005, 2008; Subramaniam et al. 2007, 2008). For example, Subramaniam et al. (2007) assessed the sensitivity of model-predicted tumor response to two major Conolly et al. assumptions. First, Conolly et al. pooled all National Toxicology Program historical controls to establish basal mutation rates. Second, data obtained by lumping animals found to have tumors at scheduled sacrifice with animals that died as a result of tumors were used to estimate formaldehyde-specific mutation rates. Subramaniam et al. also examined the choice for the first-order clearance of DPX used by Conolly et al. to match in vivo data with a slower clearance determined in vitro and how this could still be consistent with the in vivo data. The purpose of the Subramaniam et al. work was to explore whether the mutagenic mode of action, as exemplified by DPX as a precursor, has an insignificant role in predicted tumor responses. Their analysis indicated that in the range of exposures in which tumors occur, the mutagenic mode of action could have contributed up to 74% of the added tumor probability, whereas Conolly et al. (2003, 2004) showed very little contribution by this mode of action in the human when they used an upper bound on the DPX proportionality constant rather than the maximum-likelihood estimate of zero from the rat data. The committee agrees with EPA that existing data are insufficient to establish the potential biologic variability in model parameters associated with the mutagenic mode of action adequately. However, because the mutagenic mode of action is the major reason for adopting the default low-dose linear extrapolation methods over application of the BBDR models in the draft assessment, the committee recommends that the manipulations that lead to such high contributions of mutagenicity to the mode of action for nasal tumors be reconciled with the observations that formaldehyde is endogenous, that nasal tumors are very rare in both rats and humans, and that no increases in tumor frequency have been observed in animal studies at formaldehyde exposure concentrations that do not also cause cytotoxicity.

In a follow-on paper, Crump et al. (2008) evaluated the sensitivity of model output to varying such parameters as mutation, birth, and death rates of initiated cells and concluded that small changes in these parameters can result in similar fits to experimental data but yield markedly different low-dose extrapolations. The authors argued that—inasmuch as there are no data that define mutation, birth, and death rates of initiated cells, let alone data that identify what an initiated cell is—the onus is on others to demonstrate that the small changes that they made in the Conolly et al. (2004) model cannot occur. However, that can never be established with certainty. Although the committee finds that testing of alternative models and the variability in model parameters is consistent with EPA's cancer guidelines, some of the more extreme model scenarios should not have been used as a basis for rejecting the BBDR approach. In particular, adjustments of parameter values associated with mutation, birth, and death rates of initiated cells used in EPA's analysis of alternative models that yielded the most extreme deviations from the Conolly et al. (2004) low-dose extrapolations also produced unrealistically high added risks for humans at concentrations that have been observed in the environment of occupationally exposed workers (100% incidence at concentrations as low as about 0.1-1 ppm). Thus, the committee recommends that manipulations of model parameters that yield results that are biologically implausible or inconsistent with the available data be discarded and not used as a basis for rejecting the overall model.

The committee concludes that the existing BBDR models were developed from an impressive exposure- and time-dependent database on the modes of actions of formaldehyde in rats and monkeys; exquisitely detailed dosimetry models in rats, monkeys, and humans; and sparse data on humans that required scale-up of key model parameter values from animal studies and other biologic data and epidemiologic observations to constrain the human model predictions. The scope of the research makes this one of the best-developed BBDR models to date for any chemical, even with its acknowledged uncertainties. The committee also acknowledges that the draft IRIS assessment provides a thorough review of the BBDR models, the major assumptions underpinning the extrapolation to humans, and EPA's own series of papers that evaluated the sensitivity of the BBDR models to these assumptions even though the committee may not agree with the validity of all the resulting manipulations.

**Should the Biologically Based Dose-Response Models Be Used in the Environmental Protection Agency Quantitative Assessment?**

As a result of the agency's reanalysis of the models, EPA chose not to use the full rat and human BBDR models to estimate unit risks. Instead, in a benchmark-dose approach, EPA used the CFD-derived determinations of formaldehyde flux to the entire surface of mucus-coated epithelium to derive a point of departure based on nasal cancers in rats. It then extrapolated to zero dose by using a default linearized multistage approach.

The committee is concerned about that approach for low-dose extrapolation. The committee found that the evaluations of the original models and EPA's reanalysis conflicted with respect to the intent or purpose of using the formaldehyde BBDR models in human health assessments. The conflict is evident in the discussions in the draft IRIS assessment. For example, the reanalysis by Subramaniam et al. is used to support the mutagenic mode of action of formaldehyde and to reduce support for using the BBDR models on the basis of the uncertainties in parameter estimation and assumptions in the models. In contrast, Conolly et al. (2003) focused their model parameter estimates to represent "best-fit," using maximum likelihood estimates, whereas Subramaniam et al. and Crump et al. pushed parameter assumptions in a single direction to show that different assumptions that fit the experimental data can yield different results of low-dose extrapolation. The committee is concerned about the possibility that those adjustments of the Conolly et al. models may not be scientifically defensible. The committee was also struck by the relative lack of transparency in the draft IRIS assessment's description of the decision to use the peer-reviewed BBDR models minimally.

The Conolly et al. (2003, 2004) rat and human dose-response models, as the authors themselves state, are biologically motivated and mechanistic. The main purpose of a mechanistic model is to predict as accurately as possible a response to a given exposure and to provide a rational framework for extrapolations outside the range of experimental data and between species. The biologic basis of the mechanistic models implies that parameters associated with anatomy, physiology, physical and chemical properties, biochemical interactions, and dynamic responses are constrained within biologic and physical limits; they are unlike purely empirical, statistical, or mathematical models that are fit to a given dataset.

A key feature of a biologically motivated mechanistic model is that all relevant data must be reconciled with the model and that, if they cannot be, a reasonable explanation for disagreements must be articulated. Model predictions must also be reconciled with plausible outcomes, which serve as final constraints on model structure and parameter estimates. Finally, and perhaps most important, the integrated model structure forces one to identify and articulate the greatest uncertainties and variability that would affect model outcome (that is, model sensitivity) and the focus of future research. That last feature is perhaps the greatest value of the model. Although several assumptions and parameter estimates were adjusted in the Conolly et al. (2003, 2004) models in a stated attempt to make the models more conservative (that is, to imply greater risk for a given exposure or to open the door to alternative modes of action), the main use of the models was to provide best estimates of risk that could then be adjusted to incorporate uncertainties and variability in a health-protective manner. In the view of the models' developers, the adjustments of some of the model parameter values were sufficiently conservative and plausible. EPA disagreed with the contention that the Conolly et al. models were conservative and sought to evaluate and identify parameters that the models were most sensitive to and sources of uncertainty in the data and the models.

The committee acknowledges and agrees with EPA that the Conolly et al. (2003, 2004) models, like most models, contain weaknesses and tenuously supported assumptions. Conolly and co-workers felt that they made several conservative assumptions in their models— use of hockey-stick rather than J-shaped models for cell proliferation, use of overall respiratory tract cancer incidence in humans to calculate basal mutation rates, and use of an upper bound on the proportionality parameter relating DPX to mutation. EPA pushed that concept further by making even more conservative assumptions within the models that cumulatively resulted in radical departures from the results of the Conolly et al. models with regard to low-dose extrapolation of tumor incidence. The committee notes that EPA forced changes in the model parameter values in a direction that yielded more conservative results rather than one that yielded a best fit to the data.

The committee is also concerned that EPA directed substantial effort toward refuting many of the assumptions and conclusions of the Conolly et al. (2003, 2004) models rather than trying to fill the data gaps that were clearly articulated by the models. One of the major strengths of the Conolly et al. models is that the developers had to reconcile all the data or specify why they could not be reconciled. The integrated model structure allows one to identify uncertainties clearly and how they may affect model behavior. Conolly and co-workers were clear on that point and expressed the need for new data that could anchor many of the parameter values that had to be optimized from rather sparse data sets. Regardless, the committee recommends that for completeness and transparency the BBDR models published by Conolly et al. (2003, 2004), with the flaw in one numeric approach identified by EPA corrected, be used in the draft IRIS assessment and that the results be compared with those of the approach that was used in the draft assessment.

## CONCLUSIONS AND RECOMMENDATIONS

The draft IRIS assessment of formaldehyde provides an exhaustive discussion of formaldehyde toxicokinetics, carcinogenic modes of action, and various models. Although the committee agrees with much of the narrative, several issues need to be addressed in the revision of the draft assessment. First, there is broad agreement that formaldehyde is normally present in all tissues, cells, and bodily fluids and that natural occurrence complicates any formaldehyde risk assessment. Thus, an improved understanding of when exogenous formaldehyde exposure appreciably alters normal endogenous formaldehyde concentrations is needed. One approach that EPA could use would be to complete an analysis of variability and uncertainty in measuring and predicting target-tissue formaldehyde concentrations among species. Only with such an analysis can one begin to identify and address openly and transparently the question of how much added risk for an endogenous compound is acceptable.

Second, inhaled formaldehyde, a highly reactive chemical, is absorbed primarily in the upper airways and remains predominantly in the respiratory epithelium. The weight of evidence indicates that formaldehyde probably does not appear in the blood as an intact molecule except at doses high enough to overwhelm the metabolic capability of the exposed tissue. The draft IRIS assessment presents divergent opinions regarding the systemic delivery of formaldehyde that need to be resolved.

Third, the committee agrees with EPA that formaldehyde is a genotoxic chemical and that it is therefore reasonable to conclude that formaldehyde may act through a mutagenic mode of action. However, cytotoxicity and compensatory cell proliferation also appear to play important roles in formaldehyde-induced nasal tumors. Although the draft IRIS assessment discusses that mode of action, it relies on the mutagenic mode of action to justify low-dose extrapolations. The committee recommends that EPA provide alternative calculations that factor in nonlinearities associated with the cytotoxicity-compensatory cell proliferation mode of action and assess the strengths and weaknesses of each approach.

Fourth, over the last decade, several models have been developed to help to evaluate the risks associated with formaldehyde exposure, and EPA extensively evaluated several of them. EPA did use the CFD models to derive human equivalent concentrations but restricted their application to the experimental range used in the animal studies and did not extrapolate to low exposures. The committee, however, recommends that the CFD models be used to extrapolate to low concentrations, that the results be included in the overall evaluation, and that EPA explain clearly its use of the CFD modeling approaches. Furthermore, EPA, on the basis of extreme alternative model scenarios, chose not to use the BBDR models developed by Conolly et al. (2003, 2004); however, the committee questions the validity of some of these scenarios and recommends that the BBDR models developed by Conolly and co-workers be used (with the flaw in one numeric approach identified by EPA corrected), that the results be compared with those of the approach currently presented in the draft IRIS assessment, and that the strengths and weaknesses of each approach be discussed.

Fifth, in rewriting the sections of the draft IRIS assessment that pertain to the topics reviewed in this chapter, EPA should consider the implications of the most recent work. References to older studies on DNA-adduct measurements may need to be reanalyzed in light of the most recent analytic techniques that achieved superior sensitivity (for example, Lu et al. 2010). In particular, the committee finds the recent study of Lu et al. (2010) to be highly informative and the first one to distinguish clearly between exogenous and endogenous formaldehyde-induced DNA adducts. Although the study does not challenge the notion that DNA adducts play only a minor, if any, role in formaldehyde genotoxicity and carcinogenicity, compared with DNA-protein cross-links, it adds to the evidence of the inability of formaldehyde to reach distant sites. Likewise, the positive study by Wang et al. (2009) is not adequately described in the draft IRIS assessment, nor is it clear to the committee why so much emphasis is placed on the study by Craft et al. (1987).

# REFERENCES

ATSDR (Agency for Toxic Substances and Disease Registry). 1999. Toxicological Profile for Formaldehyde. U.S. Department of Health and Human Services, Public Health Services, Agency for Toxic Substances and Disease Registry, Atlanta, GA. July 1999 [online]. Available: http://www.atsdr.cdc.gov/ToxProfiles/tp1 11.pdf [accessed Jan. 5, 2011].

Bogdanffy, M.S., H.W. Randall, and K.T. Morgan. 1986. Histochemical localization of aldehyde dehydrogenase in the respiratory tract of the Fischer-344 rat. Toxicol. Appl. Pharmacol. 82(3):560-567.

Bogdanffy, M.S., P.H. Morgan, T.B.Starr, and K.T. Morgan. 1987. Binding of formaldehyde to human and rat nasal mucus and bovine serum albumin. Toxicol. Lett. 38(1-2):145-154.

Boobis, A.R., G.P. Daston, R.J. Preston, and S.S. Olin. 2009. Application of key events analysis to chemical carcinogens and noncarcinogens. Crit. Rev. Food Sci. Nutr. 49(8):690-707.

Casanova, M., H.d'A. Heck, J.I. Everitt, W.W. Harrington, Jr., and J.A. Popp. 1988. Formaldehyde concentrations in the blood of Rhesus monkeys after inhalation exposure. Food Chem. Toxicol. 26(8):715-716.

Casanova, M., D.F. Deyo, and H.d'A Heck. 1989. Covalent binding of inhaled formaldehyde to DNA in the nasal mucosa of Fischer 344 rats: Analysis of formaldehyde and DNA by high-performance liquid chromatography and provisional pharmacokinetic interpretation. Fundam. Appl. Toxicol. 12(3):397-417.

Casanova, M., K.T. Morgan, W.H. Steinhagen, J.I. Everitt, J.A. Popp, and H.d'A. Heck. 1991. Covalent binding of inhaled formaldehyde to DNA in the respiratory tract of rhesus monkeys: Pharmacokinetics, rat-to-monkey interspecies scaling, and extrapolation to man. Fundam. Appl. Toxicol. 17(2):409-428.

Casanova, M., K.T. Morgan, E.A. Gross, O.R. Moss, and H.d'A. Heck. 1994. DNA–protein cross-links and cell replication at specific sites in the nose of F344 rats exposed subchronically to formaldehyde. Fundam. Appl. Toxicol. 23(4):525-536.

Casanova, M., H.d'A. Heck, and D. Janszen. 1996. Comments on 'DNA–protein crosslinks, a biomarker of exposure to formaldehyde—in vitro and in vivo studies' by Shaham et al. [letter]. Carcinogenesis. 17(9):2097-2101.

Casanova-Schmitz, M., and H.d'A. Heck. 1983. Effects of formaldehyde exposure on the extractability of DNA from proteins in the rat nasal mucosa. Toxicol. Appl. Pharmacol. 70(1):121-132.

Casanova-Schmitz, M., T.B. Starr, and H.d'A. Heck. 1984. Differentiation between metabolic incorporation and covalent binding in the labeling of macromolecules in the rat nasal mucosa and bone marrow by inhaled [$^{14}$C]- and [$^{3}$H]formaldehyde. Toxicol. Appl. Pharmacol. 76(1):26-44.

Chang, J.C., E.A. Gross, J.A. Swenberg, and C.S. Barrow. 1983. Nasal cavity deposition, histopathology, and cell proliferation after single or repeated formaldehyde exposures in B6C3F1 mice and F-344 rats. Toxicol. Appl. Pharmacol. 68(2):161-176.

Conolly, R.B., P.D. Lilly, and J.S. Kimbell. 2000. Simulation modeling of the tissue disposition of formaldehyde to predict nasal DNA-protein cross- links in Fischer 344 rats, Rhesus monkeys, and humans. Environ. Health. Perspect. 108(suppl. 5):919-924.

Conolly, R.B., J.S. Kimbell, D.B. Janszen, and F.J. Miller. 2002. Dose response for formaldehyde-induced cytotoxicity in the human respiratory tract. Regul. Toxicol. Pharmacol. 35(1):32-43.

Conolly, R.B., J.S. Kimbell, D. Janszen, P.M. Schlosser, D. Kalisak, J. Preston, and F.J. Miller. 2003. Biologically motivated computational modeling of formaldehyde carcinogenicity in the F344 rat. Toxicol. Sci. 75(2):432-447.

Conolly, R.B., J.S. Kimbell, D. Janszen, P.M. Schlosser, D. Kalisak, J. Preston, and F.J. Miller. 2004. Human respiratory tract cancer risks of inhaled formaldehyde: Dose-response predictions derived from biologically motivated computational modeling of a combined rodent and human dataset. Toxicol. Sci. 82(1):279-296.

Costa, S., P. Coelho, C. Costa, S. Silva, O. Mayan, L.S. Santos, J. Gaspar, and J.P. Teixeira. 2008. Genotoxic damage in pathology anatomy laboratory workers exposed to formaldehyde. Toxicology 252(1-3):40-48.

Craft, T.R., E. Bermudez, and T.R. Skopek. 1987. Formaldehyde mutagenesis and formation of DNA–protein crosslinks in human lymphoblasts in vitro. Mutat. Res. 176(1):147-155.

Crump, K.S., R.P. Subramaniam, and C.B. Van Landingham. 2005. A numerical solution to the nonhomogeneous two-stage MVK model of cancer. Risk Anal. 25(4):921-926.

Crump, K.S., C. Chen, J.F. Fox, C. Van Landingham, and R. Subramaniam. 2008. Sensitivity analysis of biologically motivated model for formaldehyde-induced respiratory cancer in humans. Ann. Occup. Hyg. 52(6):481-495.

Crump, K.S., C. Chen, W.A. Chiu, T.A. Louis, C.J. Portier, R.P. Subramaniam, and P.D. White. 2010. What role for biologically based dose-response models in estimating low-dose risk? Environ. Health Perspect. 118(5):585-588.

Egle, Jr., J.L. 1972. Retention of inhaled formaldehyde, propionaldehyde, and acrolein in the dog. Arch. Environ. Health. 25(2):119-124.

EPA (U.S. Environmental Protection Agency). 1991. Formaldehyde Risk Assessment Update, Final Draft. Office of Toxic Substances, U.S. Environmental Protection Agency, Washington, DC. June 11, 1991.

EPA (U.S. Environmental Protection Agency). 2005. Guidelines for Carcinogen Risk Assessment. EPA/630/P-03/001F. Risk Assessment Forum, U.S. Environmental Protection Agency, Washington, DC. March 2005 [online]. Available: http://www.epa.gov/raf/publications/pdfs/CANCER_GUIDELINES_FINAL_3-2 5-05.PDF [accessed Nov. 23, 2010].

EPA (U.S. Environmental Protection Agency). 2010. Toxicological Review of Formaldehyde (CAS No. 50-00-0) – Inhalation Assessment: In Support of Summary Information on the Integrated Risk Information System (IRIS). External Review Draft. EPA/635/R-10/002A. U.S. Environmental Protection Agency, Washington, DC [online]. Available: http://cfpub.epa.gov/ncea/iris_drafts/recor display.cfm?deid=223614 [accessed Nov. 22, 2010].

Franks, S.J. 2005. A mathematical model for the absorption and metabolism of formaldehyde vapour by humans. Toxicol. Appl. Pharmacol. 206(3):309-320.

Garcia, G.J., J.D. Schroeter, R.A. Segal, J. Stanek, G.L. Foureman, and J.S. Kimbell. 2009. Dosimetry of nasal uptake of water-soluble and reactive gases: A first study of interhuman variability. Inhal. Toxicol. 21(7):607-618.

Georgieva, A.V., J.S. Kimbell, and P.M. Schlosser. 2003. A distributed parameter model for formaldehyde uptake and disposition in the rat nasal lining. Inhal. Toxicol. 15(14):1435-1463.

Heck, H.d'A., and M. Casanova. 1994. Nasal dosimetry of formaldehyde: Modeling site specificity and the effects of preexposure. Inhal. Toxicol. 6(suppl.):159-175.

Heck, H.d'A, and M. Casanova. 2004. The implausibility of leukemia induction by formaldehyde: A critical review of the biological evidence on distant-site toxicity. Regul. Toxicol. Pharmacol. 40(2):92-106.

Heck, H.d'A., E.L. White, and M. Casanova-Schmitz. 1982. Determination of formaldehyde in biological tissues by gas chromatography/mass spectrometry. Biomed. Mass Spectrom. 9(8):347-353.

Heck, H.d'A., M. Casanova-Schmitz, P.B. Dodd, E.N. Schachter, T.J. Witek, and T. Tosun. 1985. Formaldehyde ($CH_2O$) concentrations in the blood of humans and Fischer-344 rats exposed to $CH_2O$ under controlled conditions. Am. Ind. Hyg. Assoc. J. 46(1):1-3.

Hernandez, O., L. Rhomberg, K. Hogan, C. Siegel-Scott, D. Lai, G. Grindstaff, M. Henry, and J.A. Cotruvo. 1994. Risk assessment of formaldehyde. J. Hazard. Mater. 39(2):161-172.

Hubal, E.A., P.M. Schlosser, R.B. Conolly, and J.S. Kimbell. 1997. Comparison of inhaled formaldehyde dosimetry predictions with DNA-protein cross-link measurements in the rat nasal passages. Toxicol. Appl. Pharmacol. 143(1):47-55.

Kimbell, J.S., E.A. Gross, D.R. Joyner, M.N. Godo, and K.T. Morgan. 1993. Application of computational fluid dynamics to regional dosimetry of inhaled chemicals in the upper respiratory tract of the rat. Toxicol. Appl. Pharmacol. 121(2):253-263.

Kimbell, J.S., E.A. Gross, R.B. Richardson, R.B. Conolly, and K.T. Morgan. 1997. Correlation of regional formaldehyde flux predictions with the distribution of formaldehyde-induced squamous metaplasia in F344 rat nasal passages. Mutat. Res. 380(1-2):143-154.

Kimbell, J.S., R.P. Subramaniam, E.A. Gross, P.M. Schlosser, and K.T. Morgan. 2001a. Dosimetry modeling of inhaled formaldehyde: Comparisons of local flux predictions in the rat, monkey, and human nasal passages. Toxicol. Sci. 64(1):100-110.

Kimbell, J.S., J.H. Overton, R.P. Subramaniam, P.M. Schlosser, K.T. Morgan, R.B. Conolly, and F.J. Miller. 2001b. Dosimetry modeling of inhaled formaldehyde: Binning nasal flux predictions for quantitative risk assessment. Toxicol. Sci. 64(1):111-121.

Lu, K., L.B. Collins, H. Ru, E. Bermudez, and J.A. Swenberg. 2010. Distribution of DNA adducts caused by inhaled formaldehyde is consistent with induction of nasal carcinoma but not leukemia. Toxicol. Sci. 116(2):441-451.

Lu, K., B. Moeller, M. Doyle-Eisele, J. McDonald, and J.A. Swenberg. 2011. Molecular dosimetry of $N^2$-hydroxymethyl-dG DNA adducts in rats exposed to formaldehyde. Chem. Res. Toxicol. 24: 159-161.

Moeller, B.C., K. Lu, M. Doyle-Eisele, J. McDonald, A. Gigliotti, and J.A. Swenberg. 2011. Determination of N2-hydroxymethyl-dG adducts in the nasal epithelium and bone marrow of nonhuman primates following 13CD2-formaldehyde inhalation exposure. Chem. Res. Toxicol. 24(2):162-164.

Monticello, T.M., K.T. Morgan, J.I. Everitt, and J.A. Popp. 1989. Effects of formaldehyde gas on the respiratory tract of rhesus monkeys. Pathology and cell proliferation. Am. J. Pathol. 134(3):515-527.

Monticello, T.M., F.J. Miller, and K.T. Morgan. 1991. Regional increases in rat nasal epithelial cell proliferation following acute and subchronic inhalation of formaldehyde. Toxicol. Appl. Pharmacol. 111(3):409-421.

Monticello, T.M., J.A. Swenberg, E.A. Gross, J.R. Leininger, J.S. Kimbell, S. Seilkop, T.B. Starr, J.E. Gibson, and K.T. Morgan. 1996. Correlation of regional and nonlinear formaldehyde-induced nasal cancer with proliferating populations of cells. Cancer Res. 56(5):1012-1022.

Overton, J.H., J.S. Kimbell, and F.J. Miller. 2001. Dosimetry modeling of inhaled formaldehyde: The human respiratory tract. Toxicol. Sci. 64(1):122-134.

Patterson, D. L., E.A. Gross, M.S. Bogdanffy, and K.T. Morgan. 1986. Retention of formaldehyde gas by the nasal passages of F-344 rats. Toxicologist 6: 55.

Schlosser, P.M. 1999. Relative roles of convection and chemical reaction for the disposition of formaldehyde and ozone in nasal mucus. Inhal. Toxicol. 11(10):967-980.

Schripp, T., C. Fauck, and T. Salthammer. 2010. Interferences in the determination of formaldehyde via PTR-MS: What do we learn from m/z 31? Int. J. Mass Spectrom. 289(2-3):170-172.

Shaham, J., Y. Bomstein, A. Meltzer, Z. Kaufman, E. Palma, and J. Ribak. 1996. DNA–protein crosslinks, a biomarker of exposure to formaldehyde—in vitro and in vivo studies. Carcinogenesis. 17(1):121-125.

Shaham, J., Y. Bomstein, A. Melzer, and J. Ribak. 1997. DNA-protein crosslinks and sister chromatid exchanges as biomarkers of exposure to formaldehyde. Int. J. Occup. Environ. Health. 3(2):95-104.

Shaham, J., Y. Bomstein, R. Gurvich, M. Rashkovsky, and Z. Kaufman. 2003. DNA-protein crosslinks and p53 protein expression in relation to occupational exposure to formaldehyde. Occup. Environ. Med. 60(6):403-409.

Spanel, P., and D. Smith. 2008. Quantification of trace levels of the potential cancer biomarkers formaldehyde, acetaldehyde and propanol in breath by SIFT-MS. J. Breath Res. 2(4):1-10.

Subramaniam, R.P., K.S. Crump, C. Van Landingham, P. White, C. Chen, and P.M. Schlosser. 2007. Uncertainties in the CIIT model for formaldehyde-induced carcinogenicity in the rat: A limited sensitivity analysis-I. Risk Anal. 27(5): 1237-1254.

Subramaniam, R.P., C. Chen, K.S. Crump, D. DeVoney, J.F. Fox, C.J. Portier, P.M. Schlosser, C.M. Thompson, and P. White. 2008. Uncertainties in biologically-based modeling of formaldehyde-induced respiratory cancer risk: Identification of key issues. Risk Anal. 28(4):907-923.

Swenberg, J.A., C.S. Barrow, C.J. Boreiko, H.d'A. Heck, R.J. Levine, K.T. Morgan, and T.B. Starr. 1983. Non-linear biological responses to formaldehyde and their implications for carcinogenic risk assessment. Carcinogenesis. 4(8):945-952.

Swenberg, J.A., K. Lu, B.C. Moeller, L. Gao, P.B. Upton, J. Nakamura, and T.B. Starr. 2011. Endogenous versus exogenous DNA adducts: Their role in carcinogenesis, epidemiology and risk assessment. Toxicol Sci. 120(Supl.1):S130-S145.

Wang, M., G. Cheng, S. Balbo, S.G. Carmella, P.W. Villalta, and S.S. Hecht. 2009. Clear differences in levels of a formaldehyde-DNA adduct in leukocytes of smokers and nonsmokers. Cancer Res. 69(18):7170-7174.

Yang, Y., B.C. Allen, Y.M. Tan, K.H. Liao, and H.J. Clewell III. 2010. Bayesian analysis of a rat formaldehyde DNA-protein cross-link model. J. Toxicol. Environ. Health A. 73(12):787-806.

# 4

# Portal-of-Entry Health Effects

The Environmental Protection Agency (EPA) evaluated an array of health effects associated with formaldehyde exposure. The health effects can be characterized as portal-of-entry effects or systemic effects. The committee defined *portal-of-entry effects* as those that arise from direct interaction of inhaled formaldehyde with the airways or from the direct contact of airborne formaldehyde with the eyes or other tissue. It defined *systemic effects* as effects that occur outside those systems.

EPA's evaluation of portal-of-entry health effects—which are irritation, decreased pulmonary function, respiratory tract pathology, asthma, and respiratory tract cancers—is reviewed in this chapter. The committee determined whether EPA identified the appropriate studies, whether the studies were thoroughly evaluated, whether hazard identification was conducted appropriately according to EPA guidelines, and whether the best studies were advanced for calculation of the reference concentration (RfC) or unit risk.

For two portal-of-entry effects (irritation and decreased lung function), evidence was available from chamber studies that used brief, controlled exposures to assess acute responses and from epidemiologic studies that evaluated chronic exposures primarily in a residential setting and prevalence of symptoms or diseases or the degree of lung-function impairment. Both types of studies have strengths and weaknesses for serving as the basis of candidate RfCs. The chamber studies involve exposures to known concentrations of formaldehyde without the presence of other air contaminants, and outcome measures can be rigorously measured. However, the study populations are selected groups of volunteers, more sensitive people may avoid participation, and the numbers of participants in the studies are generally small, leading to inadequate statistical power to detect biologically significant changes in many of the studies. Furthermore, there is uncertainty about extrapolating from an acute exposure to a chronic exposure, which would be required for derivation of an RfC; and for irritant responses, observations made for a single, brief exposure may not reflect the consequences of sustained exposure. There is some indication from acute and short-terms studies that irritant responses to formaldehyde are lessened by acclimatization.

The epidemiologic studies considered in the draft IRIS assessment are primarily cross-sectional studies and subject to the general weaknesses that affect studies of this design, including the simultaneous measurement of exposure and outcome. Many of the studies involved exposure in residences, and the exposure-assessment protocols covered only a brief time window, leaving the possibility that exposures were misclassified. Furthermore, few of the studies took into account potentially confounding exposures, such as secondhand smoke or other air pollutants. The epidemiologic studies do have the advantage of assessing the risks of formaldehyde exposures as they are experienced on a chronic basis. The study populations cover the range of susceptibility and, to the extent that the effects of formaldehyde exposure are modified by interactions with other agents, the exposure to formaldehyde is experienced along with exposure to the many other contaminants in indoor air.

Given the quite different strengths and weaknesses of the two lines of research, the findings from chamber studies and epidemiologic studies should be considered as complementary. The draft IRIS assessment sets aside the chamber studies as less relevant to derivation of candidate RfCs, but the findings from the studies could be useful, and the committee does not concur with EPA's decision to set them aside. Specific recommendations are provided below for the individual health outcomes.

## IRRITATION

Formaldehyde is a reactive gas that has been consistently shown to be an eye, nose, and throat irritant. Sensory irritants act at the sensory fibers of the trigeminal nerve in the nose and upper respiratory tract. Sensory-irritation end points include self-reported sensations of pain, burning, or itching and objective measures of eye-blink counts and lacrimation (Doty et al. 2004). Although EPA's review focuses on eye, nose, and throat irritation, other types of irritation, such as dermal irritation, have been reported. EPA selected sensory irritation as a candidate critical effect on the basis of concentration-response relationships between formaldehyde and eye irritation observed in three epidemiologic studies of residential populations.

### Study Identification

EPA identified many studies that evaluated sensory irritation in response to formaldehyde exposure in residential, occupational, and clinical settings in humans and in experimental animal studies. Human studies evaluated sensory irritation responses in the eyes, nose, and throat after exposure to formaldehyde at 100-3,000 ppb and for durations ranging from 90 sec in chamber studies to chronic residential exposure. They included potentially sensitive members of the population: children less than 4 years old, adults over 65 years old, and people who have asthma.

EPA appears to have identified all appropriate exposure-response studies in humans and animals, but the literature review of studies related to the mode of action of sensory irritation associated with formaldehyde exposure should be expanded. The literature on the biologic basis of sensory irritation is more extensive than that included in the draft IRIS assessment and includes studies relevant for evaluating the mode of action of formaldehyde in the respiratory system.

### Study Evaluation

EPA summarized human and animal studies that were identified as having data on formaldehyde concentrations and sensory irritation responses in the eyes, nose, and throat. Population characteristics, exposure assessment, exposure-response relationships, and data analysis presented by the study authors were discussed by EPA. However, the committee found that study details (such as age ranges of study participants, sampling durations, and participant-selection processes) and study weaknesses (such as the limitations of the exposure assessments performed in the residential and occupational epidemiologic studies) were not thoroughly presented or critically evaluated in a consistent manner by EPA. In some cases, EPA did not give sufficient weight to study weaknesses, such as bias in the selection of participants and the possibility of confounding by other pollutants.

### Hazard Identification and Use of EPA Guidelines

Formaldehyde is a well-recognized reactive and irritant gas. EPA does not have a separate guidance document for evaluating sensory irritation responses. However, the assessment of the available human and animal studies for development of an RfC for sensory irritation was consistent with the guidance for evaluation of studies in the RfC guidelines (EPA 1994).

EPA's discussion of sensory irritation included direct sensory responses and reflex responses observed in humans and animals. The draft IRIS assessment cites the Arts et al. (2006) analysis to support its conclusion that the onset and severity of irritant responses to formaldehyde were observed to be time-dependent and concentration-dependent. However, the results of the Arts et al. (2006) analysis are incorrectly characterized and do not provide strong support for that conclusion.[1] The committee evaluated several recent reviews of formal-

---

[1]EPA (2010a) on page 5-4 states that "Arts et al. [2006] reviewed several studies and performed BMD analyses, reporting 10% extra risk BMCL values for reported eye discomfort of 560 and 240 ppb for 3 and 5 hour exposures, respectively. LOAELs of

dehyde sensory irritation and did not identify any studies explicitly designed to characterize the relationship of response, concentration, and exposure duration for sensory irritation for either acute or chronic exposures.[2] As noted by EPA, the chamber studies demonstrate that formaldehyde exposure causes sensory irritation in humans; this finding supports the association of increased sensory irritation with increased formaldehyde concentration observed in the residential epidemiologic studies. The potential contribution of sensory irritation to other respiratory health effects was acknowledged during the discussion of other effects, such as lung function, respiratory tract pathology, sensitization, and asthma.

Chapter 4.4.1 of the draft IRIS assessment provides a possible mode of action for sensory irritation: "formaldehyde-induced stimulation of the trigeminal nerve (though whether formaldehyde acts as a direct agonist is unknown)" (EPA 2010a, p. 4-458). Chapter 6 of the draft assessment makes a stronger statement about the same mode of action. Both sections, however, omit discussion of activity related to the transient receptor potential (TRP) and its association with sensory irritation. Several papers have identified the TRP family of ion channels in sensory neurons as important mediators of response to chemical irritants (Bautista et al. 2006; Macpherson et al. 2007; Bessac and Jordt 2008; Caceres et al. 2009; Bessac and Jordt 2010). Formaldehyde has been shown to activate the TRPA1 ion channel irreversibly by covalent modification—the same as the activation mechanism of other known sensory irritants, such as mustard oil and cinnamaldehyde (Macpherson et al. 2007)—and to act on TRPA1 channels to elicit pain (Macpherson et al. 2007; McNamara et al. 2007). Work by Caceres et al. (2009) provides evidence that TRPA1 plays a critical role in allergic asthmatic responses as a major neuronal mediator of allergic airway inflammation. Other environmental irritants—including the metabolites of naphthalene and styrene, ozone, acrolein, and products of lipid peroxidation resulting from oxidative stress—have been shown to activate TRPA1 (Bautista et al. 2006; Macpherson et al. 2007; Taylor-Clark et al. 2008; Taylor-Clark and Undem 2010; Lanosa et al. 2010). Multiple endogenous and exogenous agents may activate the TRPA1 ion channel simultaneously (Macpherson et al. 2007; Bessac and Jordt 2008). The committee suggests that EPA review this research and consider its potential for improving understanding of the mode of action underlying the irritant effects associated with formaldehyde exposure.

---

1,000 ppb and 1,700 ppb were reported for 1-2 minute exposures (Bender et al., 1983; Weber-Tschopp et al., 1977). These acute studies support a role for both concentration and duration in the effect level for eye irritation." However, Arts et al. (2006) use the irritation results collected after 2.5 hr of exposure, not 5 hr, for the BMD analyses because they could not get acceptable model fit using the 5-hr data.

[2]The reviews of sensory irritation by Paustenbach et al. (1997), Arts et al. (2006), and Wolkoff and Nielsen (2010) did not identify any studies in which concentration and exposure duration were systematically varied.

## Study Selection for Calculation of Reference Concentration and Identification of Point of Departure

Three epidemiologic studies—Hanrahan et al. (1984), Ritchie and Lehnen (1987), and Liu et al. (1991)—that evaluated sensory irritation in residents of mobile and conventional homes were advanced as a group by EPA and considered adequate for calculation of candidate RfCs. The studies provided concentration-response data on several sensory irritation responses, including irritation of the eyes, nose, and throat, of which eye irritation was identified by EPA as the most sensitive and best characterized. The committee agrees with EPA's decision to advance the eye irritation effects observed in the residential epidemiologic studies in spite of their limitations. However, it found that EPA set aside the chamber and occupational studies too soon in the process. Although the chamber studies are of acute duration (5-hr maximum single exposure), they are complementary with the residential studies and provide controlled measures of exposure and response. Therefore, the committee strongly recommends that EPA also present the concentration-response data from the occupational, chamber, and residential studies on the same graph and include the point estimate and measures of variability in the exposure concentrations and responses. The concentration-response relationship for eye irritation among the different types of human studies would strengthen EPA's argument for selection of residential studies for development of candidate RfCs.

The strength of the selected epidemiologic studies lies in their evaluation of responses in the general human population who are chronically exposed, their measurement of formaldehyde concentrations in residences, and their assessment of effects during or soon after sampling. EPA concluded that potential weaknesses of the studies—use of subjective surveys to collect response information, short sampling duration, and potential bias in selection of homes—were sufficiently controlled for by the study authors. However, the committee has concerns about the potential weaknesses, especially of the study conducted by Ritchie and Lehnen (1987). There are several general concerns that are relevant to each study: they are cross-sectional in design, the formaldehyde concentration measurements were taken during brief intervals and may not accurately represent usual exposure concentrations, and the investigators considered potential confounding by other pollutants to a varying extent.

The committee identified the most serious problems in the study by Ritchie and Lehnen (1987). The committee concluded that the positive attributes of that study—the large sample (2,007) and administration of the survey assessing health effects by a technician at the time of sampling—did not outweigh the potential for selection bias in self-selection of participants, who before participation in the study had to meet with a physician for prescreening and have a written request from the physician to the Minnesota Department of Health (MDH).[3]

---

[3]The committee notes that the draft assessment omitted from its description that the participants had to meet with a physician as part of the prescreening process. It states on

That process is likely to have resulted in enrichment of the sample with people who were symptomatic and concerned about formaldehyde exposure. At that time, there was substantial controversy concerning formaldehyde exposure, and people who experienced symptoms and had knowledge of formaldehyde sources in their homes would have been more likely to have sought a test from a physician and have a referral to the MDH. The very high rate (86-93%) of participants who reported eye irritation at concentrations of 300 ppb or greater, particularly in comparison with the prevalence estimates for the middle exposure category, suggests considerable participant selection bias. The draft IRIS assessment does not address that issue but comments on recall bias, noting that participants were not aware of formaldehyde concentration when the questionnaires were completed. The committee further notes that mothers responded for their children and that the analytic strategy did not account for the data structure (that is, household was the unit of assignment for exposure, but the data were analyzed as though the data from individuals within a household were independent). Thus, the committee recommends that the Ritchie and Lehnen (1987) study not be used to estimate a point of departure for a candidate RfC.

Although the contribution of cigarette smoke to sensory irritation was controlled for in the residential epidemiologic studies, the absence of evaluation of chemicals other than formaldehyde in the indoor air samples and their potential to confound the association of formaldehyde and sensory irritation is not directly addressed in the draft IRIS assessment.

EPA identified a point of departure for each study that was selected for derivation of a candidate RfC for eye irritation: a no-observed-adverse-effect level (NOAEL) of 50 ppb (Ritchie and Lehnen 1987), a lower 95% confidence limit on the benchmark concentration corresponding to a 10% response level (BMCL$_{10}$) of 70 ppb (Hanrahan et al. 1984), and a lowest observed-adverse-effect level (LOAEL) of 95 ppb (Liu et al. 1991). The committee supports the points of departure selected by EPA for the Hanrahan et al. (1984) and Liu et al. (1991) studies. Although the committee does not recommend that EPA advance the Ritchie and Lehnen (1987) study for calculating a candidate RfC, it is included in its comments on the point of departure.

The draft IRIS assessment appears to use an inconsistent approach for identifying points of departure from studies that present exposure as categories or ranges of concentrations. For example, the Ritchie and Lehnen (1987) and Liu et al. (1991) studies determined points of departure on the basis of the results of three exposure categories. Neither study had a nonexposed reference group for evaluating background response rate. Ritchie and Lehnen reported a 1-2% prevalence of eye irritation in the lowest exposure group (exposed to formaldehyde at less than 100 ppb) and a 12-32% prevalence in the middle expo-

---

page 4-2 that "in this cross-sectional study of nearly 2,000 Minnesota residents living in 397 mobile and 494 conventional homes, personal data and formaldehyde samples were collected from residents that had responded to an offer by the state health department to test homes for formaldehyde."

sure group (exposed at 100-300 ppb). EPA identified less than 100 ppb as the NOAEL and assigned 50 ppb (the midpoint between 0 ppb and 100 ppb) as the NOAEL for calculation of the candidate RfC. In the other case, Liu and co-workers reported a prevalence of eye irritation of 11-13% in the lowest exposure group (exposed at less than 70 ppb; detection limit for a 7-day passive air sample was 10 ppb) and a prevalence of 15-17% in the middle exposure group (exposed at 70-120 ppb). EPA selected 70-120 ppb as the LOAEL and 95 ppb (the midpoint of the range) as the LOAEL for calculation of the candidate RfC. The uncertainty associated with the process for selecting a point of departure is not explicitly discussed in the draft assessment.

The discussions of uncertainty associated with the points of departure for individual critical studies in Sections 5.1.4.1 and 6.2.1.4.1 of the draft IRIS assessment are too limited. For example, the draft assessment does not discuss uncertainty in the points of departure contributed by sources specific to the formaldehyde database, such as differences in methods used by the critical studies to adjust exposures (such as exposure estimated from samples collected for 7 days vs one or two sample collections of 30 or 60 min each) to reflect chronic exposure and differences in methods of characterizing exposure-response relationships (such as using regression coefficients to estimate a BMC and BMCL for a specific study's response rate vs using the midpoint of an exposure category as the estimate of the exposure concentration associated with the study's response).

## Conclusions and Recommendations

The committee agrees with EPA's selection of eye irritation as a critical sensory-irritation effect caused by formaldehyde exposure because residential, occupational, and chamber studies have demonstrated that the eyes are more sensitive to irritation from formaldehyde than the nose and throat. The committee supports EPA's advancement of the residential studies by Liu et al. (1991) and Hanrahan et al. (1984) for derivation of candidate RfCs as adequately conducted studies of a randomly selected general population and agrees with the points of departure identified by EPA from these studies:

LOAEL = 95 ppb (Liu et al. 1991)
$BMCL_{10}$ = 70 ppb (Hanrahan et al. 1984)

The committee recommends that EPA address the following in the revision of the formaldehyde draft IRIS assessment:

- Strengthen its critical evaluation of the studies.
- Re-evaluate the chamber and occupational studies for calculation of candidate RfCs.

- Not advance the Ritchie and Lehnen (1987) study for calculation of a candidate RfC.
- Review research on the TRPA1 and TRPV1 ion channels and use the information to strengthen discussion of the mode of action underlying the sensory irritation and respiratory effects associated with formaldehyde exposure.
- Add a figure that contains all the studies that evaluated eye irritation, include for each study the mean concentration, the concentration range, and the participant response rate, and organize the data by study population (residential, occupational, and chamber).

## DECREASED PULMONARY FUNCTION

Pulmonary function is assessed with spirometry, which measures the amount of air and the speed at which the air is exhaled during a forced exhalation after a maximum inhalation. Commonly used measures of pulmonary function include the total amount of air exhaled (forced vital capacity, FVC), the amount of air exhaled in the first second of exhalation (forced expiratory volume in 1 sec, $FEV_1$), the ratio of $FEV_1$ to FVC ($FEV_1$/FVC ratio), and the peak expiratory flow rate (PEFR). Pulmonary-function testing is an important tool for the assessment of both asthma and chronic obstructive pulmonary disease. The mode of action of formaldehyde's effect on pulmonary function may be direct irritation of the airways that result in an inflammatory response or in an asthmatic response in sensitive people.

### Study Identification

Acute and chronic adverse effects of occupational and residential exposures to formaldehyde on measures of pulmonary function have been investigated in several epidemiologic studies. The study populations have included occupational groups exposed to formaldehyde in various trades and industries, medical students exposed to formaldehyde in anatomy laboratories, and children and adults exposed to formaldehyde in indoor air coming from residential sources. The research approaches have included cross-sectional studies that involved testing workers' lung function before and after work shifts and cohort studies with follow-up ranging from weeks to years. The studies have used standard methods of pulmonary-function testing, including spirometry and peak-flow measurement. There is substantial literature on standardizing the testing methods and guidance on interpreting the findings (Miller et al. 2005).

Chamber studies involving brief exposures of volunteers to formaldehyde have also been conducted and have provided mixed findings. The draft IRIS assessment discusses those studies in a descriptive fashion. Deficits in pulmonary function due to formaldehyde exposure have been demonstrated in some human experimental studies that included exercise. An adverse effect of formaldehyde on pulmonary function has generally not been observed in studies of healthy volunteers who were not exercising. The lack of evidence of an effect of formaldehyde on pulmonary function in many of the chamber studies might be explained by their small samples and by the acute nature of the exposure in the experiments. Because of those issues, the chamber studies are of limited use for estimating an RfC for pulmonary-function loss. However, given the small samples, a more formal analysis that includes a display of the data with appropriate forest plots might be helpful.

EPA's review covered the relevant body of epidemiologic and experimental literature. The evidence is diverse and comes from multiple exposed populations. Many of the studies were performed several decades ago and reflect the substantial public-health concern at the time. The review does not appear to have missed more recent studies.

## Study Evaluation

EPA's review of epidemiologic studies, toxicologic studies, and experimental-chamber studies of formaldehyde and pulmonary function was thorough and appropriate. Here, EPA used tables to summarize the pulmonary function studies; this aided the committee in its review of the information.

## Hazard Identification and Use of EPA Guidelines

EPA concluded that there is extensive evidence that formaldehyde causes decreased pulmonary function in humans (Section 6.1.3.9, EPA 2010a). Although the committee agrees with that conclusion, a clear narrative is needed to provide the rationale for it. There are no specific EPA guidelines for evaluating effects of agents on pulmonary function or other respiratory disease outcomes. The most relevant may be EPA's RfC guidelines (EPA 1994). Inconsistencies in the approach taken by EPA may reflect the lack of adequate guidance for this domain of health outcomes.

## Study Selection for Calculation of Reference Concentration and Identification of Point of Departure

EPA selected the findings in children (6-15 years old) from the Arizona study by Krzyzanowski et al. (1990) as the basis for the development of a candidate RfC for decreased pulmonary function as measured by PEFR. The draft IRIS assessment justifies the choice by stating that "the best single study demonstrating decreased pulmonary function is the moderate residential study by Krzyzanowski et al. (1990)" (EPA 2010a, pp. 5-36 to 5-39). The draft discusses only briefly the possibility of using other studies, such as the Kriebel et al. (1993, 2001) studies of anatomy students exposed to formalin. The committee notes that the Krzyzanowski et al. (1990) findings are inherently limited by the cross-sectional nature of their study and found that the study design is not sufficiently described in the published report.

Krzyzanowski et al. (1990) found an effect of formaldehyde in children but not in adults. The findings from the studies by Kriebel et al. (1993, 2001) of anatomy students indicate that the effects of formaldehyde on pulmonary function in adults may be more severe in asthmatics. EPA should provide a more thorough analysis and rationale for its choice to advance only the Krzyzanowski et al. (1990) study but also consider the Kriebel et al. (1993, 2001) studies as additional candidates for its assessment.

EPA derived a $BMCL_{10}$ of 17 ppb as the point of departure on the basis of the study by Krzyzanowski et al. (1990). Regression coefficients were estimated by using a linear mixed-effects regression model, presented in Table 5 in Krzyzanowski et al. (1990), and used by EPA to derive a $BMCL_{10}$. The PEFR model allowed the effect of formaldehyde exposure to depend on time of day (morning vs bedtime) and asthmatic status. The calculation of a $BMCL_{10}$ implies the estimation of the dose associated with a particular response level. The model predicts that the dose will vary with the presence of effect modifiers (morning exposure and asthma) of the exposure of interest (formaldehyde). The draft IRIS assessment is unclear about how EPA defined the $BMCL_{10}$ given the effect modification. Greater elaboration and discussion of how a $BMCL_{10}$ was based on the model fit are needed.

## Conclusions and Recommendations

EPA's review and evaluation of the literature on the effects of formaldehyde on pulmonary function were thorough and appropriate. Although the committee supports EPA's determination that exposure to formaldehyde may cause a decrease in pulmonary function, EPA should provide a stronger narrative to support that conclusion. The committee agrees with the choice of the Krzyzanowski et al. (1990) study as the basis of the derivation of a point of departure for a candidate RfC but recommends that other studies also be considered for calculation of an alternative point of departure.

The committee recommends that EPA address the following in the revision of the formaldehyde draft IRIS assessment:

- Prepare plots of the findings of the chamber studies to assess the utility of pooling their results.
- Provide further justification for its choice of the study by Krzyzanowski et al. (1990) for estimating the point of departure.
- Consider estimating an alternative point of departure based on the studies by Kriebel et al. (1993, 2001).
- Provide a clear description of how the data from the study by Krzyzanowski et al. (1990) were used to estimate the $BMCL_{10}$.

## NONCANCER RESPIRATORY TRACT PATHOLOGY

Formaldehyde-induced effects on the respiratory tract have been studied extensively. Animal studies have clearly shown that inhaled formaldehyde at 2 ppm or higher is cytotoxic and that increases in epithelial cell proliferation occur after chronic formaldehyde inhalation by mice, rats, and nonhuman primates (Kerns et al. 1983; Monticello et al. 1996). The resulting airway lesions include rhinitis, epithelial dysplasia, and squamous metaplasia. Formaldehyde-induced effects on the respiratory tract demonstrate concentration, time, and site dependence, and these lesions exhibit an anterior to posterior severity gradient (Kerns et al. 1983; Monticello et al. 1996). The committee concludes that the effects for which a candidate RfC should be calculated are histopathologic lesions of the nasal epithelium.

### Study Identification

The draft IRIS assessment reviews six studies that examined the effects of formaldehyde exposure on the human upper respiratory tract. Two that examined the same worker cohort are identified as the most robust and sensitive and are selected for possible derivation of a candidate RfC (Holmstrom and Wilhelmsson 1988; Holmstrom et al. 1989). The draft assessment also reviews the extensive literature on histopathologic effects in the respiratory tract and effects on mucociliary clearance in laboratory animals that inhaled formaldehyde. There are numerous studies in several species of laboratory animals, including ones using acute, subchronic, and chronic inhalation exposures.

Although the committee did not perform its own literature search, it notes that two papers (Schoenberg and Mitchell 1975; Bracken et al. 1985) directly related to formaldehyde exposure and cited by Holmstrom and Wilhelmsson (1988) are not included in the draft IRIS assessment. Despite that oversight, EPA appears to have identified the appropriate animal and human studies.

### Study Evaluation

The review of the literature in the draft IRIS assessment is extensive but is often unfocused and lacks critical evaluation of the studies. The animal studies are presented in detail, and they provide unequivocal evidence that inhalation of formaldehyde by laboratory animals causes histopathologic lesions of the upper respiratory tract. The six studies that evaluated formaldehyde-induced effects in humans all used relatively small samples, and the methods of characterizing exposure were not always discussed. None of the human studies demonstrated that exposure duration was important or that a concentration-response relationship was present. The draft assessment appears to give equal weight to all publications of the human and animal studies, and there is no consideration of study quality, of the validity of the measurement of the exposure concentration, or of whether a study was conducted under good laboratory practices or their equivalent. A critical analysis of the human study selected for possible derivation of a candidate RfC is lacking, and there is no evidence of a specified format for evaluation of the animal or human studies. The committee concludes that those are all important weaknesses of the draft assessment.

### Hazard Identification and Use of EPA Guidelines

EPA has no specific guidelines for evaluating the pathology of the respiratory tract. However, the animal studies provide clear evidence that inhaled formaldehyde causes lesions of the nasal epithelium. Specifically, it causes cell death that is followed by regenerative hyperplasia and metaplasia of the epithelium of the upper respiratory tract (Monticello et al. 1996). Those responses are probably the combined result of overloading of host protective mechanisms, such as mucociliary clearance, detoxification, and DNA repair (Kerns et al. 1983). The draft IRIS assessment provides a summary of respiratory toxicity (see pages 4-467 through 4-469) that adds little to the previous detailed discussion of the human and animal studies provided elsewhere. The mode of action for the development of histopathologic lesions of the respiratory tract in that section is not clearly presented.

The draft IRIS assessment concludes that histopathologic lesions and abnormal mucociliary clearance are equivalent pathologic lesions of the upper respiratory tract. The committee agrees with the statement that "the mucociliary apparatus is an important barrier to infection and exogenous agents and, thus, [a change in mucociliary clearance] is considered as a potential adverse effect" (EPA 2010a, p. 4-67). The committee, however, concludes that the data on animals and humans are not consistent enough to support derivation of a point of departure for mucociliary clearance and that abnormal mucociliary clearance is not equivalent to a histopathologic lesion.

### Study Selection for Calculation of Reference Concentration and Identification of Point of Departure

The epidemiologic study (Holmstrom and Wilhelmsson 1988; Holmstrom et al. 1989) that was selected for advancement was the strongest of the six available that studied histopathology of the human upper respiratory tract. However, as discussed in detail below, even that study had substantial weaknesses that limit its use for derivation of a point of departure and calculation of a candidate RfC. In contrast, numerous studies of several species of animals could be used to derive a candidate RfC. The committee recommends that EPA use the animal data to calculate a candidate RfC for respiratory tract lesions in the revised IRIS assessment. That would provide a basis for evaluating the uncertainty associated with the other candidate RfCs that have been calculated.

The human study selected for advancement (reported in two publications) involved 70 workers exposed to formaldehyde in a chemical plant that produced formaldehyde for resins and impregnation of paper for laminate production (Holmstrom and Wilhelmsson 1988; Holmstrom et al. 1989). The study included a second group of 100 workers exposed to wood dust and formaldehyde in a furniture-production facility. The reference group consisted of 36 persons, most of whom were office workers in the same village as the furniture workers. The draft IRIS assessment does not adequately report the exposure concentrations that were reported in the publications and does not adequately discuss the cohort exposed to wood dust and formaldehyde. The mean formaldehyde exposure of the group exposed only to formaldehyde was accurately reported by EPA as 0.210 ppm. However, EPA did not report the exposure range (0.040-0.403 ppm) or the frequent peak short-term exposures (up to 0.810 ppm) provided by the study authors. Thus, there was considerable variability in the exposures that occurred in the occupational study that would not be reflected by the mean exposure data. In addition, 31 of the 70 formaldehyde-exposed workers were potentially exposed to paper dust at up to 1 $mg/m^3$; this exposure is not noted or discussed in the draft IRIS assessment despite being reported by the study authors, and the committee is concerned that the coexposure could be a confounding factor in the study. The furniture workers were exposed to formaldehyde at 0.160-0.243 ppm (mean, 0.202 ppm) and to wood dust at 1-2 $mg/m^3$. The formaldehyde exposure of the office workers was measured at 0.073-0.137 ppm in late summer. The mean exposure was 0.073 ppm, on the basis of four measurements in different seasons. The background formaldehyde exposure of the reference group is not mentioned in the draft IRIS assessment. The group exposed to formaldehyde alone had an increased nasal-resistance score (as measured by rhinomanometry) of 10.2 compared with 6.5 in the reference group; the difference was not statistically significant. The nasal mucociliary clearance was delayed in both exposure groups compared with the reference group but the difference was statistically significant ($p < 0.05$) only in the formaldehyde group. Participants who were identified with "pathologically slow nasal clearance" amounted to 14 of 69 in the formaldehyde group and 14 of 95 in the formaldehyde and wood group compared with one of 36 in the reference group. That difference may have been statistically significant, but its biologic significance has not been established. Moreover, the outcome was not found to be related to exposure concentration or duration when subgroups were examined. EPA concluded that the study demonstrated a LOAEL of 0.210 ppm on the basis of impairment of mucociliary clearance. On the basis of the available study data, 0.073 ppm (the background concentration for the reference group) might represent a NOAEL.

Many of the subjects in the study also had nasal biopsies taken and evaluated for histopathologic lesions on an 8-point scale in which 0 was normal epithelium, 4 was stratified squamous epithelium with marked horny layer, and 8 was carcinoma. The formaldehyde-exposed group had a mean score of 2.16, the wood dust and formaldehyde group had a mean score of 2.07, and the control group had a mean score of 1.56. The difference was significant ($p < 0.05$) only in the formaldehyde group. However, the actual exposure of the 62 members of the formaldehyde group examined histopathologically was reported as 0.240 ppm rather than the 0.210 ppm reported for mucociliary clearance. The higher value was selected by EPA as the point of departure. It is the opinion of the committee that this study has numerous weaknesses, the most important of which is a failure to identify a clear relationship between adverse responses and exposure concentration or exposure duration.

Table 5-4 of the draft IRIS assessment lists the point of departure as 240 ppb as a LOAEL for the upper respiratory tract pathology. As noted above, that is the value reported by Holmstrom et al. (1989) for histopathologic lesions on the basis of nasal biopsy specimens. However, EPA's analysis does not account for a background exposure of 73 ppb in the reference group (a possible NOAEL). Section 5.1.2.1 goes on to note that the study did not report a concentration-response relationship and states that "this is less exact than other available studies which provide exposure-response relationships" (EPA 2010a, p. 5-36). It is also noted that animal studies support sensory irritation as a more sensitive end point than histopathologic changes in the nasal mucosa. Therefore, a candidate RfC for that end point was not calculated. The committee agrees that this study should not be used for calculation of a candidate RfC. However, the committee does not agree that there are sufficient data from animal studies to support the conclusion that sensory irritation is a more sensitive end point than histopathologic changes in the nasal mucosa. One form of sensory irritation that has been studied in animals and reviewed in the draft IRIS assessment is reflex bradypnea, a response that occurs in mice and rats exposed to formaldehyde vapors. The concentration that causes an acute 50% reduction in the respiratory rate has been reported to be about 3 ppm in mice (Kane and Alarie 1977) and about 10 ppm in rats (Cassee et al. 1996). In contrast, rats exposed to formaldehyde at 2 ppm for 24 months reportedly have respiratory tract lesions (Kerns et al. 1983). Those data suggest that sensory irritant responses seen in animals occur at concentrations associated with airway histopathologic changes.

As noted above, numerous well-documented studies have reported the occurrence of upper respiratory tract pathology in laboratory animals, including nonhuman primates, after inhalation of formaldehyde. The dataset is one of the most extensive available on an inhaled chemical. The computational-fluid-dynamics (CFD) models could be used to predict the dose to the human upper respiratory tract and thereby decrease the need for uncertainty factors associated with animal-to-human extrapolation. The CFD models have been used by EPA for cancer risk assessment and would be applicable to respiratory tract toxicity (see Chapter 3 for a discussion of the CFD models). As stated previously, the committee recommends that EPA use the animal data to calculate an RfC for respiratory tract pathology.

## Conclusions and Recommendations

The committee concludes that a candidate RfC should be calculated for noncancer pathology of the respiratory tract (that is, in the nasal epithelium). The committee agrees with EPA that the human studies that include that end point are not sufficiently complete and that they should not be used to calculate a candidate RfC. However, many well-conducted animal studies have reported noncancer histopathologic lesions of the respiratory tract after formaldehyde inhalation.

The committee recommends that EPA address the following in the revision of the formaldehyde draft IRIS assessment:

- Calculate a candidate RfC from the animal data.
- Do not calculate a candidate RfC for mucociliary clearance.

## ASTHMA

*Asthma* refers to a broad phenotype of respiratory disease that may differ by age and by causal agent (NHLBI 2007). In the United States, over 22 million people have asthma (NHLBI 2007). Asthma is the most common chronic disease of childhood and affects about 8-10% of children in the United States and a similar percentage of adults (EPA 2010b). Clinically, the phenotype of asthma involves intermittent airflow obstruction with wheezing and shortness of breath, although in some people the airflow obstruction may become persistent. The term is applied to a broad array of phenotypes involving wheezing illnesses, some of which do not correspond directly with the strict definition of asthma. In infants and young children, wheezing illnesses that are the result of lower respiratory tract infections (such as respiratory syncytial virus) are often labeled as asthma (Martinez et al. 1995). Follow-up of children who have had early-life wheezing illnesses has shown that many did not have incident asthma as the cause of the clinical picture. Similarly, in adults, the phenotypes of asthma and chronic obstructive pulmonary disease may overlap (Barnes 2008).

The underlying pathogenetic process involves inflammation of the airways, which occurs in response to multiple environmental triggers (NHLBI 2007; Barnes 2008). The disease aggregates in families, so it is considered to have a genetic basis, although the search for key underlying genes has been difficult and not yet definitive. The committee notes that an environmental agent like formaldehyde potentially could increase the incidence of asthma, lead to a more severe clinical phenotype, or alter the natural history, perhaps by sustaining inflammation.

### Study Identification

As acknowledged in the draft IRIS assessment, formaldehyde might have effects on various components of the asthma phenotype: allergic sensitization, incidence of asthma, the prevalence of asthma, the severity of asthma, increased symptoms and exacerbations, and lower lung function. The committee concurs with the focus on those aspects of asthma occurrence and of the clinical phenotype. However, given the scope of the outcomes of interest, the document should provide the search terms used to ensure that all relevant literature was identified. A broad set of studies are potentially relevant to this set of outcome measures, and consideration is given to the various studies relevant to each phenotype component.

The committee could not identify major studies that were not included but notes that most of the studies considered were completed several decades ago, when there was substantial interest in formaldehyde as an indoor air pollutant. When the studies were conducted, however, the asthma phenotype was not nearly as well characterized as it is now, so although the selected studies were considered by EPA to address asthma, the phenotypic characterizations in children are subject to misclassification when viewed in the context of current understanding. The committee was particularly concerned about the phenotype considered in the study by Rumchev et al. (2002).

## Study Evaluation

The relevant studies were observational and, to a lesser extent, experimental. The principal concerns regarding the experimental studies that involved exposure of asthmatic volunteers to formaldehyde are participant characteristics, particularly the severity of their asthma, which affects the magnitude of response and generalizability of findings. In connection with the observational studies, a broader set of methodologic concerns that should have been systematically reviewed can be identified, such as the basis of the characterization of phenotype and the exposure estimation strategy.

The review of the studies is largely descriptive and without a specified format for study evaluation. In the case of selected studies, particularly those considered to be more informative, some attention is given to methodologic concerns but in a nonsystematic fashion. The key limitation of cross-sectional data is mentioned but the implications are not sufficiently explored. The key issue of asthma phenotype was not addressed in the draft assessment. Consequently, the committee found the review of studies to be inadequate.

## Hazard Identification and Use of EPA Guidelines

The hazard identification discussion (EPA 2010a, pp. 4-462 to 4-467) is largely descriptive and repeats the previous descriptions of the studies, capturing some of the main features and findings. There is also a lack of clarity concerning the health end point considered—that is, incidence, prevalence, or exacerbation of established asthma. The evidence from epidemiologic studies is summarized as follows: "in conclusion, the epidemiologic studies of formaldehyde exposure among children support the finding that low indoor and outdoor concentrations result in increased asthma incidence and prevalence" (EPA 2010a, p. 4-466). The conclusion is reached that the mode of action has "not been elucidated," although several studies are indicated as lending "weight of evidence to a neurogenic" mode of action. Much of the discussion of mechanisms is speculative and unreferenced. For example, the draft assessment states that "formaldehyde-induced inflammation of the airways may contribute to observed decreases in measures of pulmonary function. Even short-term inflammatory reactions could reduce the effective diameter of the conductive airways, resulting in lower respiratory volumes in a number of functional tests. Formaldehyde-induced trigeminal nerve stimulation contributes to airway inflammation, which in turn would reduce airway function" (EPA 2010a, p. 4-462). Given the extensive literature on the pathogenesis of asthma, the discussion is inadequate, and any consideration of mode of action should have been better framed in current understanding of the underlying mechanisms of asthma causation and exacerbation. The proposal concerning trigeminal nerve stimulation is not referenced, even though it receives further treatment in the second paragraph of the two that address mode of action. The discussion does not reflect the state of knowledge of asthma pathogenesis. Abundant research and review articles are available and should have been cited (see, for example, Cohn et al. 2004 and Barnes 2008).

Hazard identification is not explicitly based on a guidance document of the agency; the most relevant may be EPA's RfC guidelines (EPA 1994). The ad hoc approach taken in the draft IRIS assessment may reflect inadequate guidance on asthma. Given the limited discussion of the evidence and the lack of clear criteria for evidence evaluation, the committee did not find sufficient support for the hazard identification.

## Study Selection for Calculation of Reference Concentration and Identification of Point of Departure

Few studies were available to evaluate asthma critically; in fact, only the case-control study by Rumchev et al. (2002) addressed the diagnosis of childhood asthma, and the cross-sectional study by Garrett et al. (1999) was one of the few to include assessment of allergen sensitivity. Both included some in-home measurements of formaldehyde concentration, a strong component of the rationale for advancing the studies.

Both studies were labeled as having data of "high quality," although limitations of both were evident to the committee. The exposure protocols for both included measurements over a short period that may not reflect the biologically relevant period of exposure. The study by Rumchev et al. (2002), although interpreted as addressing the diagnosis and incidence of asthma, involved participants in an age range during which lower respiratory illnesses with wheezing that have an infectious etiology are frequently mislabeled as asthma (Martinez et al. 1995). Consequently, the relevance of this study specifically to childhood asthma should be questioned because current understanding of wheezing illnesses in this age range indicates that they are transitory and not likely to represent the onset of asthma. A later report by Rumchev et al. (2004) describes higher concentrations of a number of volatile organic compounds in the homes of the cases compared with those of the controls. The potential for confounding by those other pollutants in assessing the effect of formaldehyde was not addressed.

Garrett et al. (1999) carried out a cross-sectional study that was inherently limited with respect to causal inference and establishing the temporality of associations. The cross-sectional findings of the study are used for three outcome measures: prevalence of atopy, prevalence of asthma, and a respiratory symptom score. A study strength is inclusion of four 4-day measurement periods for formaldehyde, but the analyses are cross-sectional. The symptom data, which represented symptoms that occurred during the last year, were collected at one time.

The draft text related to selection of the two studies (EPA 2010a, p. 5-39) does not compare their characteristics and strengths and weaknesses in relation to others that were not selected. The committee finds that the array of studies considered was too narrow and that an expedient choice was made with little additional explanation for the choice. The study by Rumchev et al. (2002) concerned an outcome other than incident asthma and should not have been advanced. The study by Garrett et al. (1999) provides relevant information in spite of its cross-sectional nature. EPA has advanced other cross-sectional studies (for example, Krzyzanowski et al. 1990); thus, that limitation alone does not constitute a sufficient reason not to advance a study for RfC calculation. Consequently, the committee concurs that Garrett et al. can be advanced in the absence of more informative prospective studies.

Two studies are advanced for derivation of candidate RfCs: one for diagnosis of incident asthma based on Rumchev et al. (2002) and the other for allergic sensitization—including critical effects of allergic sensitization, asthma, and respiratory symptoms—based on Garrett et al. (1999). For diagnosis of asthma, two points of departure—a NOAEL of 33 ppb based on Rumchev et al. (2002) and a LOAEL of 28 ppb based on Garrett et al. (1999)—are provided. The committee notes that the Rumchev et al. (2002) study was omitted from Table 5-4 in the draft IRIS assessment that summarizes all studies advanced for candidate RfCs. As noted above, however the study by Rumchev et al. (2002) is inappropriately advanced.

For the Rumchev et al. (2002) study, the committee notes that a key decision made in the calculation is not well documented. Figure 4-1 (Figure 5-5 in EPA 2010a) provides the odds ratios according to measured concentrations of formaldehyde in the residences. Only the odds ratio for the highest category is statistically significant. The NOAEL is taken to be 30-49 $\mu g/m^3$ with the stated rationale that "the next highest exposure category was considered to be part of an exposure-related trend of increasing asthma risk and, therefore, biologically significant" (EPA 2010a, p. 5-46). Although confidence intervals around the two intermediate concentration points overlap, a linear model with a continuous exposure measure was statistically significant. Regardless, criteria for identifying "an exposure-related trend" are not given.

The candidate RfC values derived from the study by Garrett et al. (1999) are based on EPA's interpretation of the trends observed in the categorical analyses for the three outcomes (allergic sensitization, asthma, and respiratory symptoms). The paper provides outcome data by three levels of highest exposure measured (less than 20, 20-50, and greater than 50 ppb). There appears to be an error in the description of the categories: it refers to a middle exposure range of 16-40 ppb and a high category of greater than 40 ppb. Regardless, the approach used in the draft IRIS assessment for identifying a LOAEL is to use the midpoint of the middle exposure category. That approach is not specifically justified and appears to represent a pragmatic attempt to handle data that are provided in the form of an exposure-response relationship without a nonexposed group and no clear NOAEL.

## Conclusions and Recommendations

Asthma is a complex phenotype on whose pathogenesis substantial research has been conducted. The discussion of asthma needs to be strengthened to reflect the extensive literature better. The discussion of mode of action needs to be greatly strengthened and grounded in current understanding of pathogenesis. The current speculative discussion is not satisfactory. In light of the current understanding of wheezing illnesses in early life, the study by Rumchev et al. (2002) cannot be advanced as reflecting "asthma." The committee agrees that the study by Garrett et al. (1999) can be advanced for calculation of a candidate RfC, but the approach taken for identification of a LOAEL needs better justification.

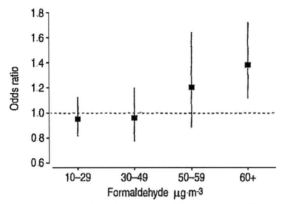

**FIGURE 4-1** Odds ratios for physician-diagnosed asthma in children associated with in-home formaldehyde concentrations in air. This is Figure 5-5 in EPA 2010a. Source: Rumchev et al. 2002. Reprinted with permission; copyright 2002, European Respiratory Society.

The committee recommends that EPA address the following in the revision of the formaldehyde draft IRIS assessment:

- Strengthen the discussion of asthma to reflect current understanding of this complex phenotype and its pathogenesis better. There should be greater clarity regarding the outcomes considered: incident asthma (the occurrence of new cases), prevalent asthma (the presence of asthma at the time of study), or exacerbation of established asthma.
- Strengthen the discussion of mode of action and cite the extensive relevant literature.
- Do not advance the study by Rumchev et al. (2002) as pertaining to asthma. That study appears relevant not to the asthma phenotype but rather to early-life wheezing illness.
- Develop better the approach taken to identifying the LOAEL in the Garret et al. (1999) study.

## RESPIRATORY TRACT CANCERS

The respiratory tract has been considered to include plausible locations for formaldehyde-induced cancer in humans because it is a site of first contact and because of the observed increased incidence of nasal tumors in laboratory animals exposed to formaldehyde. It is particularly true for cancers of the nose and nasal cavity (International Classification of Diseases, Revision 8 [ICD8] 160) and nasopharynx (ICD8 147) because the dose of formaldehyde is expected to be the greatest at these upper respiratory sites. In contrast, lung cancer (ICD8 162) is a less plausible site because the delivered dose for formaldehyde is expected to be much less in the lower respiratory tract than in the upper respiratory tract. There is extensive evidence from human and experimental studies that the mode of action of the induction of upper respiratory tract tumors by formaldehyde involves a genotoxic mechanism. It is also likely that the mode of action involves induction of cell proliferation by formaldehyde toxicity. Chapter 3 presents a more extensive discussion of the mode of action of formaldehyde induction of respiratory cancers.

### Study Identification

The draft IRIS assessment appears to have identified all the pertinent studies of formaldehyde and respiratory cancers available at the time of its release. Thus, the draft assessment presents and discusses findings on lung cancer, nasopharyngeal cancer, nasal cancer, and other respiratory cancers from a large number of occupational cohort studies and several population-based case-control studies of adults. The committee is aware that an update of the National Cancer Institute (NCI) cohort for solid tumors is in progress, and the formaldehyde IRIS assessment will need to include the update when it becomes available. However, the committee is not recommending that EPA wait until release of the update to complete its assessment.

## Study Evaluation

The draft assessment presents an extensive evaluation of the pertinent studies published before EPA's review. Particular attention was appropriately given to discussion of the findings of an excess of nasopharyngeal cancer (NPC) in the NCI study of workers employed in formaldehyde industries (Hauptmann et al. 2004). The draft assessment also considers alternative analyses and criticisms of the study by Marsh and colleagues (Marsh et al. 1996, 2002, 2007a,b; Marsh and Youk 2005). A primary finding of the NCI study is that the excess of NPC in the cohort was attributable almost entirely to one of the 10 study facilities (Marsh et al. 2007a). That facility was one of the largest; this would translate into greater statistical power to detect an increase in NPC mortality than some of the smaller facilities, which had fewer cases of this very rare cancer and thus much lower statistical power. The draft assessment suggests that there was also an excess of NPC in a second plant, but this finding was too unstable to be useful given that it was based on only one case and had a very wide confidence interval (standardized mortality ratio, 5.35; 95% confidence interval, 0.13-29.83). The pooling of results across plants translates into even greater power to detect a formaldehyde-associated excess of NPC. Plant differences other than statistical power may explain differences in observed cancer rates and are worth noting as limitations in interpreting risk estimates based on this study. For example, a reanalysis of the NCI cohort by Marsh et al. (2007b) provides evidence that the excess of NPC might be explained by other employment in silver-smithing or other metal-working industries in Connecticut. However, there is no evidence from other studies that those industries are associated with an increased risk of NPC.

The study evaluation would be improved substantially if the EPA framework for causal determinations were stated explicitly because it would help to provide structure for the study evaluations and clarify which findings were most relevant to future causal determinations. The committee finds that the review regarding all the other respiratory cancer sites is thorough and appropriate.

## Hazard Identification and Use of EPA Guidelines

The draft IRIS assessment draws inconsistent conclusions about hazard identification in four sections. In the summary of the section "Respiratory Tract Cancer," the draft concludes that there is sufficient epidemiologic evidence that formaldehyde is causally associated with NPC and sinonasal cancer (EPA 2010a, section 4.1.2.1.5.4.). In the section "Summary: Carcinogenic Hazards in Humans," the draft concludes that "the weight of the epidemiologic evidence at this time supports a link between formaldehyde exposure and NPC in humans" but does not mention sinonasal cancer (EPA 2010a, section 4.1.2.3., p. 4-188). In the section "Synthesis and Evaluation of Carcinogenicity: Cancers of the Respiratory Tract," the text offers the conclusion that "formaldehyde is causally related to cancers of the upper respiratory tract as a group" (EPA 2010a, section 4.5.1., p. 4-486). In the section "Hazard Characterization for Formaldehyde Carcinogenicity," the draft states that "human epidemiological evidence is sufficient to conclude a causal association between formaldehyde exposure and nasopharyngeal cancer, nasal and paranasal cancer" (EPA 2010a, section 4.5.4., p. 4-535). The inconsistencies in the conclusions obviously need to be resolved.

Because the draft IRIS assessment presents no causal framework explicitly, the committee considered the appropriateness of EPA's conclusions in the context of EPA's *Guidelines for Carcinogen Risk Assessment* (EPA 2005). The guidelines state that for a substance to be a known human carcinogen, there should be "convincing epidemiologic evidence of a causal association between human exposure and cancer" or, exceptionally, if all the following conditions are met: "(a) there is strong evidence of an association between human exposure and either cancer or the key precursor events of the agent's mode of action but not enough for a causal association, and (b) there is extensive evidence of carcinogenicity in animals, and (c) the mode(s) of carcinogenic action and associated key precursor events have been identified in animals, and (d) there is strong evidence that the key precursor events that precede the cancer response in animals are anticipated to occur in humans and progress to tumors, based on available biological information" (EPA 2005, p. 2-54).

EPA's conclusion that NPC is causally related to formaldehyde was based on the positive findings of the NCI cohort study (Hauptmann et al. 2004) and on several case-control studies (Olsen et al. 1984; Rousch et al. 1987; West et al. 1993; Vaughan et al. 2000). Although an excess of NPC was not observed in other cohort studies of formaldehyde-exposed workers (Coggon et al. 2003; Pinkerton et al. 2004), the negative findings might be explained by the low statistical power of the studies for these rare tumors. There was a consensus in the committee that it would be consistent with EPA guidelines to draw a causal conclusion for NPC and formaldehyde on the basis of the combination of the epidemiologic findings with experimental data and mechanistic data on formaldehyde.

For nasal and paranasal (that is, sinonasal) cancers, EPA found a causal relationship with formaldehyde exposure on the basis of three factors: positive findings of several case-control studies (Hayes et al. 1986; Olsen and Asnaes 1986; Luce et al. 2002), stronger associations between formaldehyde and cancer in a neighboring tissue (NPC), and an excess of nasal cancer in rats exposed to formaldehyde (EPA 2010a). The committee concluded that EPA's causal determination regarding sinonasal cancer is consistent with its cancer guidelines.

In Section 4.5.1 of the draft IRIS assessment (EPA 2010a), EPA extends its determination of causality to include all upper respiratory cancers and formaldehyde. EPA does not define what it means by "all upper respiratory cancers," but it might be taken to include cancers of the oral cavity and larynx, as well as nasopharyngeal and sinonasal cancers. That determination was made even though little evidence about any upper respiratory cancer site other than NPC or sinonasal cancer was offered. The committee does not find that determination to be consistent with EPA's cancer guidelines.

The draft IRIS assessment does not conclude that there is a causal association between exposure to formaldehyde and cancers of the lung. Only one of the major cohort studies of formaldehyde-exposed workers has reported a significant excess of lung cancer (Coggon et al. 2003). In addition, studies of deposition of formaldehyde in the respiratory tract have demonstrated clearly that the amount of formaldehyde deposited in the lower respiratory tract would be low. The committee concurs with EPA that there is a lack of sufficient evidence of an increased risk of lung cancer in humans exposed to formaldehyde.

## Study Selection for Calculation of Cancer Unit Risk

EPA selected the study by Hauptmann et al. (2004) as the basis of its exposure-response assessment of formaldehyde and NPC. That was the only possible choice because the Hauptmann et al. (2004) study was the only study from which an exposure-response relationship could be derived. Furthermore, that study has a number of strengths for conducting an exposure-response assessment. In addition to having individual estimates of formaldehyde exposure, the study had a long period of follow-up, controlled for a number of potential confounding variables, and used internal comparisons that minimized biases related to the healthy-worker effect.

The strengths, however, are offset by a number of weaknesses. First, as noted earlier, the excess of NPC in the study was due to an excess in only one of the 10 study facilities. Although that pattern may be explained by the low statistical power of the individual study sites for this rare cancer, it raises concerns about the generalizability of the findings to the other facilities and to other workers exposed to formaldehyde. It also raises the possibility that the results were confounded by other pollutants present at that one facility. Second, the exposure-response findings for NPC were far less significant for cumulative formaldehyde exposure ($p = 0.032$) than for peak exposures ($p < 0.001$). Despite that finding, EPA chose to construct a dose-response relationship by using findings on cumulative exposure, given that peak exposures could not be used to estimate risks associated with lifetime exposures to environmental concentrations of formaldehyde. Finally, a serious concern has been raised about an unexplained under ascertainment of deaths in the Hauptmann et al. (2004) study (Marsh et al. 2010). In the update of findings on lymphatic and hematopoietic cancers in the NCI cohort, Beane-Freeman et al. (2009) noted that 1,006 deaths that had occurred before 1995 were missing from earlier analyses of the NCI cohort. The Beane-Freeman analysis also extended follow-up of the cohort to 2004. The additional follow-up period resulted in a total of 7,091 additional deaths, which is nearly double the number that were included in Hauptmann et al. (2004). The effect of that under ascertainment of deaths and the additional follow-up period has important implications for analyses of the NCI cohort and NPC. Given the importance of the NCI study to the formaldehyde assessment, EPA should make an effort to update its assessment once the NCI study findings on NPC become available.

EPA also conducted an exposure-response analysis based on the combined findings of an increase in nasal squamous cell carcinoma in two long-term bioassays in F344 rats (Kerns et al. 1983; Monticello et al. 1996). The results of the exposure-response models were used to estimate risk of nasal cancer and of cancer of the entire respiratory tract (upper and lower) in humans. The committee concurs with EPA's decision to use the data as the basis of the development of a model to estimate risk of nasal cancer but has serious doubts about the appropriateness of using the study to estimate risks of all respiratory cancers in humans. Those doubts and the lack of evidence from most of the epidemiologic studies of an excess of respiratory cancers other than sinonasal cancer and NPC support the committee's conclusion that risks of all respiratory cancers (particularly lower respiratory tract cancers) should not be calculated now.

### Conclusions and Recommendations

EPA's review of the literature on formaldehyde and respiratory cancer was thorough and appropriate. It would be useful if, in the future, EPA could explicitly state its criteria for evaluation of the evidence of causality based on its own cancer guidelines. Several sections of the draft IRIS assessment contain conflicting statements on the evidence of causality that clearly need to be rectified. The committee finds that, on the basis of EPA's guidelines, there is sufficient evidence of a causal association between formaldehyde and cancers of the nose and nasal cavity (ICD8 160) and nasopharynx (ICD8 147) but not other sites of respiratory tract cancer. The committee agrees that the study by Hauptmann et al. (2004) is an appropriate choice for the derivation of a point of departure and unit risk. Although it is a high-quality study, it is important to recognize some of its deficiencies, such as the apparent inconsistency between the findings in different plants in the study and the weakness of the exposure-response relationship in connection with cumulative exposure. Furthermore, the study was found to be missing deaths in a later update of the cohort for lymphatic and hematopoietic cancers. NCI is updating its cohort for respiratory cancer and other solid tumors. The update not only will include the missing deaths but will extend the follow-up, and this will result in nearly twice the amount of deaths.

The committee recommends that EPA address the following in the revision of the formaldehyde draft IRIS assessment:

- Revise the document to state clearly the criteria that EPA used to determine the causality for cancer.

- Resolve the conflicting statements in the document concerning which upper respiratory cancer sites were found to be causally associated with formaldehyde exposure.
- Update the dose-response analysis in the IRIS assessment when the findings from the update of the NCI cohort on solid cancers become available. However, the committee is not recommending that EPA wait until release of the update to complete its assessment.

## REFERENCES

Arts, J.H., M.A. Rennen, and C. de Heer. 2006. Inhaled formaldehyde: Evaluation of sensory irritation in relation to carcinogenicity. Regul. Toxicol. Pharmacol. 44(2):144-160.

Barnes, P.J. 2008. Immunology of asthma and chronic obstructive pulmonary disease. Nat. Rev. Immunol. 8(3):183-192.

Bautista, D.M., S.E. Jordt, T. Nikai, P.R. Tsuruda, A.J. Read, J. Poblete, E.N. Yamoah, A.I. Basbaum, and D. Julius. 2006. TRPA1 mediates the inflammatory actions of environmental irritants and proalgesic agents. Cell 124(6):1269-1282.

Bessac, B.F., and S.E. Jordt. 2008. Breathtaking TRP channels: TRPA1 and TRPV1 in airway chemosensation and reflex control. Physiology 23(6):360-370.

Bessac, B.F., and S.E. Jordt. 2010. Sensory detection and responses to toxic gases: Mechanisms, health effects, and countermeasures. Proc. M. Thorac. Soc. 7(4):269-277.

Beane-Freeman, L.E., A. Blair, J.H. Lubin, P.A. Stewart, R.B. Hayes, R.N. Hoover, and M. Hauptmann. 2009. Mortality from lymphohematopoietic malignancies among workers in formaldehyde industries: The National Cancer Institute cohort. J. Natl. Cancer Inst. 101(10):751-761.

Bender, JR, L.S. Mullin, G.J. Graepel, and W.E. Wilson. 1983. Eye irritation response of humans to formaldehyde. Am. Ind. Hyg. Assoc. J. 44:463-465.

Bracken, M.J., D.J. Leasa, and W.K. Morgan. 1985. Exposure to formaldehyde: Relationship to respiratory symptoms and function. Can. J. Public Health 76(5):312-316.

Caceres, A.I., M. Brackmann, M.D. Elia, B.F. Bessac, D. del Camino, M. D'Amours, J.S. Witek, C.M. Fanger, J.A. Chong, N.J. Hayward, R.J. Homer, L. Cohn, X. Huang, M.M. Moran, and S.E. Jordt. 2009. A sensory neuronal ion channel essential for airway inflammation and hyperreactivity in asthma. Proc. Nat. Acad. Sci. 106(22):9099-9104.

Cassee, F.R., J.H. Arts, J.P. Groten, and V.J. Feron. 1996. Sensory irritation to mixtures of formaldehyde, acrolein, and acetaldehyde in rats. Arch. Toxicol. 70(6):329-337.

Coggon, D., E.C. Harris, J. Poole, and K.T. Palmer. 2003. Extended follow-up of a cohort of British chemical workers exposed to formaldehyde. J. Natl. Cancer Inst. 95(21):1608-1615.

Cohn, L., J.A. Elias, and G.L. Chupp. 2004. Asthma: Mechanisms of disease persistence and progression. Annu. Rev. Immunol. 22:789-815.

Doty, R.L., J.E. Cometto-Muniz, A.A. Jalowayski, P. Dalton, M. Kendal-Reed, and M. Hodgson. 2004. Assessment of upper respiratory tract and ocular irritative effects of volatile chemicals in humans. Crit. Rev. Toxicol. 34(2):85-142.

EPA (U.S. Environmental Protection Agency). 1994. Methods for Derivation of Inhalation Reference Concentrations and Application of Inhalation Dosimetry. EPA/600/8-

90/066F. Office of Health and Environment Assessment, Office of Research and Development, U.S. Environmental Protection Agency, Research Triangle Park, NC. October 1994 [online]. Available: http://www.epa.gov/raf/publications/pdfs/RFCMET HODOLOGY.PDF [accessed Nov. 28, 2010].

EPA (U.S. Environmental Protection Agency). 2005. Guidelines for Carcinogen Risk Assessment. EPA/630/P-03/001F. Risk Assessment Forum, U.S. Environmental Protection Agency, Washington, DC. March 2005 [online]. Available: http://www.epa. gov/raf/publications/pdfs/CANCER_GUIDELINES_FINAL_3-25-05.PDF [accessed Nov. 24, 2010].

EPA (U.S. Environmental Protection Agency). 2010a. Toxicological Review of Formaldehyde (CAS No. 50-00-0) – Inhalation Assessment: In Support of Summary Information on the Integrated Risk Information System (IRIS). External Review Draft. EPA/635/R-10/002A. U.S. Environmental Protection Agency, Washington, DC [online]. Available: http://cfpub.epa.gov/ncea/iris_drafts/recordisplay.cfm?dei d=223614 [accessed Nov. 22, 2010].

EPA (U.S. Environmental Protection Agency). 2010b. Asthma Prevalence. U.S. Environmental Protection Agency [online]. Available:
http://cfpub.epa.gov/eroe/index.cfm?fuseaction=detail.viewInd&lv=list.listByAlpha&r=2 19646&subtop=381 [accessed Nov. 23, 2010].

Garrett, M.H., M.A. Hooper, B.M. Hooper, P.R. Rayment, and M.J. Abramson. 1999. Increased risk of allergy in children due to formaldehyde exposure in homes. Allergy 54(4):330-337 [Erratum-Allergy 54(12):1327].

Hanrahan, L.P., K.A. Dally, H.A. Anderson, M.S. Kanarek, and J. Rankin. 1984. Formaldehyde vapor in mobile homes: A cross sectional survey of concentrations and irritant effects. Am. J. Public Health 74(9):1026-1027.

Hauptmann, M., J.H. Lubin, P.A. Stewart, R.B. Hayes, and A. Blair. 2004. Mortality from solid cancers among workers in formaldehyde industries. Am. J. Epidemiol. 159(12):1117-1130.

Hayes, R.B., J.W. Raatgever, A. de Bruyn, and M. Gerin. 1986. Cancer of the nasal cavity and paranasal sinuses, and formaldehyde exposure. Int. J. Cancer 37(4):487-492.

Holmstom, M., and B. Wilhelmsson. 1988. Respiratory symptoms and pathophysiological effects of occupational exposure to formaldehyde and wood dust. Scand. J. Work Environ. Health 14(5):306-311.

Holmstrom, M., B. Wilhelmsson, H. Hellquist, and G. Rosen. 1989. Histological changes in the nasal mucosa in persons occupationally exposed to formaldehyde alone and in combination with wood dust. Acta Otolayngal. 107(1-2):120-129.

Kane, L.E., and Y. Alarie. 1977. Sensory irritation to formaldehyde and acrolein during single and repeated exposures in mice. Am. Ind. Hyg. Assoc. J. 38(10):509-522.

Kerns, W.D., K.L. Pavkov, D.J. Donofrio, E.J. Gralla, and J.A. Swenberg. 1983. Carcinogenicity of formaldehyde in rats and mice after long-term inhalation exposure. Cancer Res. 43(9):4382-4392.

Kriebel, D., S.R. Sama, and B. Cocanour. 1993. Reversible pulmonary responses to formaldehyde. A study of clinical anatomy students. Am. Rev. Respir. Dis. 148(6 Pt. 1):1509-1515.

Kriebel, D., D. Myers, M. Cheng, S. Woskie, and B. Cocanour. 2001. Short-term effects of formaldehyde on peak expiratory flow and irritant symptoms. Arch. Environ. Health 56(1):11-18.

Krzyzanowski, M., J.J. Quackenboss, and M.D. Lebowitz. 1990. Chronic respiratory effects of indoor formaldehyde exposure. Environ. Res. 52(2):117-125.

Lanosa, M.J., D.N. Willis, S. Jordt, and J.B. Morris. 2010. Role of metabolic activation and the TRPA1 receptor in the sensory irritation response to styrene and naphthalene. Toxicol. Sci. 115(2):589-595.

Liu, K.S., F.Y. Huang, S.B. Hayward, J. Wesolowski, and K. Sexton. 1991. Irritant effects of formaldehyde exposure in mobile homes. Environ. Health Perspect. 94:91-94.

Luce, D., A. Leclerc, D. Begin, P.A. Demers, M. Gerin, E. Orlowski, M. Kogevinas, S. Belli, I. Bugel, U. Bolm-Audorff, L.A. Brinton, P. Comba, L. Hardell, R.B. Hayes, C. Magnani, E. Merler, S. Preston-Martin, T.L. Vaughan, W. Zheng, and P. Boffetta. 2002. Sinonasal cancer and occupational exposures: A pooled analysis of 12 case-control studies. Cancer Causes Control 13(2):147-157.

Macpherson L.J., B. Xiao, K.Y. Kwan, M.J. Petrus, A.E. Dubin, S. Hwang, B. Cravatt, D.P. Corey, and A. Patapoutian. 2007. An ion channel essential for sensing chemical damage. J. Neurosci. 27(42):11412–11415.

Marsh, G.M., and A.O. Youk. 2005. Reevaluation of mortality risks for nasopharyngeal cancer in the formaldehyde cohort study of the National Cancer Institute. Regul. Toxicol. Pharmacol. 42(3):275-283.

Marsh, G.M., R.A. Stone, N.A. Esmen, V.L. Henderson, and K.Y. Lee. 1996. Mortality among chemical workers in a factory where formaldehyde was used. Occup. Environ. Med. 53(9):613-627.

Marsh, G.M., A.O. Youk, J.M. Buchanich, L.D. Cassidy, L.J. Lucas, N.A. Esmen, and I.M. Gathuru. 2002. Pharyngeal cancer mortality among chemical plant workers exposed to formaldehyde. Toxicol. Ind. Health 18(6):257-268.

Marsh, G.M., A.O. Youk, J.M. Buchanich, S. Erdal, and N.A. Esmen. 2007a. Work in the metal industry and nasophyngeal cancer mortality among formaldehyde-exposed workers. Regul. Toxicol. Pharmacol. 48(3):308-319.

Marsh, G.M., A.O. Youk, and P. Morfeld. 2007b. Mis-specified and non-robust mortality risk models for nasophyngeal cancer in the National Cancer Institute formaldehyde worker cohort study. Regul. Toxicol. Pharmacol. 47(1):59-67.

Marsh, G.M., A.O. Youk, P. Morfeld, J.J. Collins, and J.M. Symons. 2010. Incomplete follow-up in the National Cancer Institute's formaldehyde worker study and the impact on subsequent reanalyses and causal evaluations. Regul Toxicol Pharmacol 58(2):233-236.

Martinez, F.D., A.L. Wright, L.M. Taussig, C.J. Holberg, M. Halonen, and W.J. Morgan. 1995. Asthma and wheezing in the first six years of life. The Group Health medical Associates. N. Engl. J. Med. 332(3):133-138.

McNamara, C.R., J. Mandel-Brehm, D.M. Bautista, J. Siemens, K.L. Deranian, M. Zhao, N.J. Hayward, J.A. Chong, D. Julius, M.M. Moran, and C.M. Fanger. 2007. TRPA1 mediates formalin-induced pain. Proc. Nat. Acad. Sci. 104(33):13525-13530.

Miller, M.R., R. Crapo, J. Hankinson, V. Brusasco, G. Burgos, R. Casaburi, A. Coates, P. Enright, C.P. van der Grinten, P. Gustaffson, R. Jensen, D.C. Johnson, N. MacIntyre, R. McKay, D. Navajas, O.F. Pedersen, R. Pellegrino, G. Viegi, and J. Wagner. 2005. General considerations for lung function testing. Eur. Respir. J. 26(1):153-161.

Monticello, T.M., J.A. Swenberg, E.A. Gross, J.R. Leininger, J.S. Kimbell, S. Seilkop, T.B. Starr, J.E. Gibson, and K.T. Morgan. 1996. Correlation of regional and nonlinear formaldehyde-induced nasal cancer with proliferating populations of cells. Cancer Res. 56(5):1012-1022.

NHLBI (National Heart Lung and Blood Institute). 2007. Expert Panel Report 3 (EPR3): Guidelines for the Diagnosis and Management of Asthma. U.S. Department of Health and Human Services, National Institute of Health, National Heart Lung and Blood Institute [online]. Available: http://www.nhlbi.nih.gov/guidelines/asthma/asthgdln.htm [accessed Nov. 23, 2010].

Olsen, J.H., and S. Asnaes. 1986. Formaldehyde and the risk of squamous cell carcinoma of the sinonasal cavities. Br. J. Ind. Med 43(11):769-774.

Olsen, J.H., S.P. Jensen, M. Hink, K. Faurbo, N.O. Breum, and O.M. Jensen. 1984. Occupational formaldehyde exposure and increased nasal cancer risk in man. Int. J. Cancer 34(5):639-644.

Paustenbach, D., Y. Alarie, T. Kulle, N. Schachter, R. Smith, J. Swenberg, H. Witschi, and S.B. Horowitz. 1997. A recommended occupational exposure limit for formaldehyde based on irritation. J. Toxicol. Environ. Health A 50(3):217-263.

Pinkerton, L.E., M.J. Hein, and L.T. Stayner. 2004. Mortality among a cohort of garment workers exposed to formaldehyde: An update. Occup. Environ. Med. 61(3):193-200.

Ritchie, I.M., and R.G. Lehnen. 1987. Formaldehyde-related health complaints of residents living in mobile and conventional homes. Am. J. Public Health 77(3):323-328.

Rousch, G.C., J. Walrath, L.T. Stayner, S.A. Kaplan, J.T. Flannery, and A. Blair. 1987. Nasopharyngeal cancer, sinonasal cancer, and occupations related to formaldehyde: A case-control study. J. Natl. Cancer Inst. 79(6):1221-1224.

Rumchev, K.B., J.T. Spickett, M.K. Bulsara, M.R. Phillips, and S.M. Stick. 2002. Domestic exposure to formaldehyde significantly increases the risk of asthma in young children. Eur. Respir. J. 20(2):403-408.

Rumchev, K, J. Spickett, M. Bulsara, M. Phillips, and S. Stick. 2004. Association of domestic exposure to volatile organic compounds with asthma in young children. Thorax 59(9):746-751.

Schoenberg, J.B., and C.A. Mitchell. 1975. Airway disease caused by phenolic (phenol-formaldehyde) exposure. Arch. Environ. Health 30(12):574-577.

Taylor-Clark, T.E., and B.J. Undem. 2010. Ozone activates airway nerves via the selective stimulation of TRPA1 ion channels. J. Physiol. 588(3):423-433.

Taylor-Clark, T.E., M.A. McAlexander, C. Nassenstein, S.A. Sheardown, S. Wilson, J. Thornton, M.J. Carr, and B.J. Undem. 2008. Relative contributions of TRPA1 and TRPV1 channels in the activation of vagal bronchopulmonary C-fibers by the endogenous autacoid 4-oxononenal. J. Physiol. 586(14):3447-3459.

Vaughan, T.L., P.A. Stewart, K. Teschke, C.F. Lynch, G.M. Swanson, J.L. Lyon, and M. Berwick. 2000. Occupational exposure to formaldehyde and wood dust and nasopharyngeal carcinoma. Occup. Environ. Med. 57(6):376-384.

Weber-Tschopp, A, T. Fischer, and E. Grandjean, E. 1977. Irritating effects of formaldehyde on man. Int. Arch. Occup. Environ. Health. 39(4):207-218.

West, S., A. Hildesheim, and M. Dosemeci. 1993. Non-viral risk factors for nasopharyngeal carcinoma in the Philippines: Results from a case-control study. Int. J. Cancer 55(5):722-727.

Wolkoff, P., and G.D. Nielsen. 2010. Non-cancer effects of formaldehyde and relevance for setting an indoor air guideline. Environ. Int. 36(7):788-799.

# 5

# Systemic Health Effects

As noted in Chapter 4, the health effects of exposure to formaldehyde evaluated by the Environmental Protection Agency (EPA) can be characterized as portal-of-entry effects or systemic effects. In this chapter, the committee reviews EPA's evaluation of systemic health effects, including immunotoxicity, neurotoxicity, reproductive and developmental toxicity, and lymphohematopoietic cancers. The committee determined whether EPA identified the appropriate studies, whether the studies were thoroughly evaluated, whether hazard identification was conducted appropriately in light of EPA guidelines, and whether the best studies were advanced for calculation of the reference concentration (RfC) or unit risk.

Chapter 3 of the present report addresses the question of systemic bioavailability of inhaled formaldehyde. High reactivity and extensive nasal absorption of formaldehyde restrict systemic delivery of inhaled formaldehyde beyond the upper respiratory tract and major conducting airways of the lung. Indeed, the weight of evidence suggests that it is unlikely for formaldehyde to appear in the blood as an intact molecule, except perhaps when exposure doses are high enough to overwhelm the metabolic capability of the tissue at the site of exposure. Thus, systemic responses are unlikely to arise from the direct delivery of formaldehyde (or its hydrated form methanediol) to a distant site in the body. However, it is important to distinguish between systemic delivery of formaldehyde and systemic effects. The possibility remains that systemic delivery of formaldehyde is not a prerequisite for some of the reported systemic effects seen after formaldehyde exposure. Those effects may result from indirect modes of action associated with local effects, such as irritation, inflammation, and stress. Therefore, the committee reviewed EPA's evaluation of the systemic effects and determined whether the evidence presented supported EPA's conclusions.

## IMMUNOTOXICITY

Immunomodulation or immunotoxicity occurs when environmental factors (such as stress, health status, and chemical exposure) change the homeostatic processes that regulate the immune system in susceptible populations. The consequences of immunotoxicity can be highly divergent and depend on the environmental factor, the duration and timing of the exposure, the overall health status of the exposed person, and the route of exposure. Immunotoxicity may occur from direct effects on immune cells or from indirect effects on various cell components, such as altered endocrine function. It may also occur at an anatomic site that is distant from the point of entry. Indeed, the systemic nature of the immune system may mean that an exposure at one site causes damaged or modified cells to move to another location in the body where they may mediate the effects of the toxicant.

Adverse health effects associated with immunotoxicity may include higher infection rate; alterations in lymphocyte cell populations; hyperactivity of immune cells, such as increased respiratory activity (increased production of reactive oxygen and reactive nitrogen species) and cytokine production; autoimmunity; altered immune-cell trafficking throughout the body; increased allergy or atopy; and susceptibility to cancer. Research to determine the immunotoxicity of an agent focuses on those and other responses. EPA has developed a health-effects test guideline for immunotoxicity (EPA 1998a).

In the case of formaldehyde, as has been discussed elsewhere in the present report, most of or all the direct effects occur at the point of entry in the upper respiratory tract. Immune cells in the bronchial and nasal associated lymphoid tissue (BALT and NALT) are most likely proximal targets of formaldehyde. Understanding the potential immunotoxicity of formaldehyde is therefore of critical importance.

Adverse effects of formaldehyde on BALT and NALT may be manifested systemically because these lymphoid cells migrate to the lymph nodes, spleen, liver, peripheral blood, and other immune tissues. Specifically, most BALT and NALT cells belong to the arm of the immune system referred to as the innate immune system. The role of the innate immune cells is to recognize and respond to tissue damage, apoptotic cells, and evolutionarily conserved protein and glycoprotein patterns expressed on bacteria, viruses, parasites, and other pathogens. The consequences of innate immune-cell recognition of pathogen-associated molecular patterns (PAMPs) are to increase production of reactive oxygen species, to engulf the particles expressing PAMPs, and to migrate systemically where the potential infectious agent is presented to the adaptive arm of the immune system. The role of the adaptive arm of the immune system is to produce cytokines, which activate antibody production, increase inflammation, and recruit lymphocytes to the site of infection.

Consequently, the systemic nature of the immune system and interplay between the innate and adaptive arms of the immune system suggest a plausible potential target of formaldehyde despite its limited distribution beyond the point of entry. Moreover, potential alterations of the innate immune cells mediated by formaldehyde may have profound effects on the adaptive and peripheral immune system. The draft IRIS assessment summarizes a number of human and animal studies that describe formaldehyde-induced immunotoxicity. Although many of the appropriate studies were identified, a more careful evaluation of the relative strengths and weaknesses of the key studies should have been provided. Moreover, additional weight could have been given to animal studies in which exposure assessment was more rigorously controlled and a diverse set of end points was examined. The committee recognizes that differences exist in leukemia sensitivities between animals and humans. However, the differences in responses may not be relevant for other immunotoxicities, such as respiratory burst activity, sensitivity, and atopy.

## Study Identification

The draft IRIS assessment discusses immunologic end points affected by formaldehyde exposure on the basis of human and animal studies in the compiled database of published studies. The committee did not perform an additional literature search, but it appears that, in general, the appropriate studies were identified and adequately discussed. Specifically, the draft IRIS assessment presents studies designed to address the following questions:

- Does formaldehyde exposure increase upper respiratory tract infections?
- Does formaldehyde induce lymphocyte associated respiratory burst activity and inflammation?
- Is formaldehyde exposure associated with allergic sensitivity or atopy?
- What is the toxicologic significance of antibody responses directed against formaldehyde or formaldehyde-protein complexes?

The structure of the draft assessment is such that the research questions regarding human and animal end points are addressed separately. As will be discussed below, the committee finds that a more integrated approach in which the human and animal studies of a given immunologic end point are discussed and evaluated together would result in a more concise and transparent report.

## Study Evaluation

In addressing the questions above, the draft IRIS assessment presents numerous studies that provide data that suggest that formaldehyde is immunomodulatory (EPA 2010). Specifically, in addressing the question of what effect formaldehyde has on susceptibility to upper respiratory tract infections, the draft cites Holness and Nethercott (1989), Krzyzanowski et al. (1990), and Lyapina et al. (2004). The Holness and Nethercott (1989) and Lyapina et al. (2004) studies were conducted in occupational settings, and the Kyzyzanowski et al. (1990) study was conducted in homes in which formaldehyde concentrations were measured. The concentrations in the occupational settings were 0.71-1.55 ppm, and the average concentration in the home study was 26 ppb. As reported in the draft assessment, all three studies showed an association between formaldehyde exposure and increased incidence of upper respiratory tract infections. No directly comparable animal studies that would have strengthened or weakened those findings are cited. However, formaldehyde exposure at higher exposure concentrations (2 ppm or higher) has been shown to reduce mucociliary apparatus function in the rodent (Morgan et al. 1986). Similar effects, such as slowed mucociliary clearance, have been seen in occupationally exposed people (Holmström and Wilhelmsson 1988). The mucociliary apparatus is an important barrier to infection and other exogenous agents, so the finding is supportive of the human studies. The three key studies used by EPA reflect the state of the science with respect to formaldehyde and virally induced upper respiratory tract infections. Given the small number of studies, this section of the draft IRIS assessment would have been greatly improved by a critical evaluation of those studies.

Regarding the question of whether formaldehyde affects lymphocyte respiratory burst activity or inflammation, several studies in humans and animals are listed. Specifically, Lyapina et al. (2004) used flow cytometry to measure changes in neutrophil respiratory burst activity in occupationally exposed workers who had chronic bronchitis. A weakness of the study is that the assay used to measure respiratory burst activity is not specific. Moreover, the details of the study preclude concluding whether formaldehyde exposure or a chronic bronchial condition in the selected subjects was the cause of the changed cellular activity. However, Dean et al. (1984) and Adams et al. (1987) performed animal studies using 3-week exposures to formaldehyde at 15 ppm and measured changes in peritoneal macrophage hydrogen peroxide production. As shown in Table 4-47 of the draft IRIS assessment, peroxide production was increased in response to macrophage activation. If the results of the studies were synthesized, EPA could strengthen its conclusion that formaldehyde exposure affects respiratory burst activity in the immune system. Moreover, the animal studies demonstrate effects on the innate immune system at a distant site (for example, the peritoneum); this lends credence to the biologic plausibility of systemic effects caused by formaldehyde exposure.

Studies covering sensitivity and atopy are similarly listed and described in the draft IRIS assessment with little evaluation of the strengths and weaknesses of the cited work. Moreover, in the section describing human studies, a substantial amount of text is devoted to assessing whether exposed people generate IgE antibodies against formaldehyde. IgE antibodies are generated against allergic agents and chemical haptens (chemicals complexed with endogenous proteins that elicit an immune response). Although discussion of IgE antibodies against formaldehyde is not a trivial question, this section in the draft assessment could be condensed. Furthermore, an additional question to ask in the section "Sensitivity and Atopy" would be, Does exposure to formaldehyde modulate responses to known allergens, such as dust mites, ragweed, animal dander, and mold spores? If there has been research on that subject, it is not presented in the draft assessment with respect to human exposures.

In comparison, the results of many studies in animal models support a conclusion that formaldehyde exposure modifies allergic responses. Specifically, Tarkowski and Gorski (1995), Riedel et al. (1996), and Lino dos Santos Franco et al. (2009) found increased sensitivity in rodents that were coexposed to a model allergen (ovalbumin) and formaldehyde. Sadakane et al. (2002) and Ohtsuka et al. (2003) found changes in inflammatory cytokine production in the lungs after formaldehyde exposure. Several other studies summarized in Table 4-54 of the draft IRIS assessment showed more modest results or no effect of formaldehyde. The disparate observations may be due partly to the use of different rodent species and strains in the studies. Moreover, the exposure protocols varied widely. In some cases, animals were pre-exposed to formaldehyde and then sensitized; in other studies, sensitization occurred before formaldehyde exposure; and in others, sensitization and exposure occurred simultaneously. Although the committee agrees that each type of protocol appropriately replicates a real-world exposure scenario, the section deserves a robust rubric to evaluate the strengths and weaknesses of the studies presented. That is particularly important given that the section "Sensitization and Atopy" of the draft IRIS assessment concludes with a statement that "taken as a whole, the results support the finding that formaldehyde exposure can aggravate a type I hypersensitivity response" (EPA 2010, p. 4-335). On the basis of the review currently provided, the committee cannot agree with that conclusion because no clear framework for drawing it is presented.

## Hazard Identification and Use of EPA Guidelines

Hazard identification for immunotoxicity was conducted and reported in a generally appropriate fashion, given EPA guidelines. However, the language used in the review of some studies could be improved. For example, the discussion of Riedel et al. (1996), which documented airway sensitivity in guinea pigs in response to formaldehyde at 0.13 or 0.25 ppm, uses the term *biologically significant* (EPA 2010, p. 4-319). The term is used in the absence of a statement of a statistically significant effect by the study authors. Thus, EPA should provide a justification for its conclusion that the effect was biologically significant and indicate whether additional statistical analyses were performed.

In addition, the immune-hazard identification section could have been clearer with a discussion summarizing immune effects of formaldehyde. Specifically, consistencies between human and animal findings regarding inflammation, target-cell types, and airway responses should be noted. Cells of the innate immune system appear to be targets or mediators of formaldehyde-induced immunotoxicity in animal and human studies, and a concluding statement containing that information would be useful.

### Study Selection for Calculation of Reference Concentration and Identification of Point of Departure

The sections on immunotoxicity do not identify any studies for deriving a candidate RfC. Thus, no candidate RfC was calculated for the immunotoxic effects of formaldehyde.

### Conclusions and Recommendations

The systemic nature of the immune system and the interplay between the innate and adaptive arms of the immune system provide a plausible potential target of formaldehyde, despite its limited distribution beyond the point of entry. The draft IRIS assessment summarizes many human and animal studies that describe formaldehyde-induced immunotoxicity. The committee agrees with EPA's decision not to calculate a candidate RfC for immunotoxicity at this time.

The committee recommends, however, that EPA address the following in the revision of the formaldehyde draft IRIS assessment:

- Provide a more careful evaluation of the relative strengths and weaknesses of the key studies.
- Consider giving additional weight to animal studies in which exposure assessment was more rigorously controlled.

### NEUROTOXICITY

*Neurotoxicity* is defined as any adverse effect on the chemistry, structure, or function of the nervous system during development or in maturity. Neurotoxicity may be permanent or reversible, and it can be expressed as neuropathologic effects or as neurochemical, electrophysiologic, or behavioral changes. In general, chemical-induced changes in the structure or persistent behavioral, neurochemical, or neurophysiologic changes in the nervous system are regarded as neurotoxic effects. Reversible effects occurring at doses that could endanger performance in the workplace or that are associated with a known neurotoxicologic mode of action are also considered adverse. Formaldehyde exposure via inhalation has been shown to affect nervous-system function adversely in laboratory animals and humans, although there are few data on formaldehyde-induced neurologic effects in humans.

## Study Identification

EPA appears to have identified all available relevant literature on formaldehyde neurotoxicology; the committee could not identify any important studies that were not included. The draft IRIS assessment identifies seven neurotoxicity studies that are considered as candidates for RfC development, most notably the epidemiologic studies by Weisskopf et al. (2009) and Kilburn et al. (1985, 1987). Several experimental rat studies are also identified as candidate studies for RfC development. Experimental mouse studies are noted, but they are dismissed because of confounding issues. All studies addressed exposures of short duration, so information regarding the relationship between formaldehyde toxicity and exposure duration is sparse.

## Study Evaluation

The evaluation of the epidemiologic studies in the draft IRIS assessment focused on Weisskopf et al. (2009) and Kilburn et al. (1985, 1987). Weisskopf et al. (2009) reported a statistically significant association (relative risk, 2.47; 95% CI, 1.58-3.86) between self-reported years of formaldehyde exposure and death from amyotrophic lateral sclerosis (ALS). The draft assessment concludes that the study supports the "causal association of neuropathological effects in humans following long-term formaldehyde exposure" (EPA 2010, p. 4-476). The committee, however, is not convinced that the study established a causal association. EPA's conclusion of causality between formaldehyde inhalation and the development of ALS is premature, is supported by an isolated study with limited exposure data, and lacks sufficient evidence of biologic plausibility. Indeed, the study authors stated that "the increased risk attributed to formaldehyde could be the result of exposure to some other unmeasured factor commonly associated with formaldehyde" (Weisskopf et al. 2009). Kilburn et al. (1985) reported that a group of 76 female histology technicians displayed statistically significantly greater frequencies of lack of concentration and loss of memory, disturbed sleep, impaired balance, variations in mood, and irritability than did a control group of 56 unexposed female clerical workers. The technicians had been employed for 2-37 years (mean, 12.8 years). Analysis of workplace air samples indicated the presence of several solvents, including formaldehyde (0.2-1.9 ppm), xylene (3.2-102 ppm), chloroform (2-19.1 ppm), and toluene (8.9-12.6 ppm). Thus, exposure to xylene and other solvents most likely contributed to the observed neurobehavioral effects. Kilburn (1994) also reported that three anatomists and one railroad worker, occupationally exposed to airborne formaldehyde for 14-30 years, showed impaired performance on choice reaction time, abnormal balance, digit symbol, and perceptual motor speed.

EPA's review of the candidate animal studies is largely descriptive and lacked a systematic or specified format for study evaluation. None of the candidate studies adhered to EPA's neurotoxicity-testing guidelines (EPA 1998b).

Several studies used designs that deviated substantially from the testing guidelines and common practice. In particular, animal studies performed by Malek and co-workers (2003a) used extremely short-duration (3-min) motor-activity test sessions. EPA neurotoxicity-test guidelines explicitly state that the test session should be "of sufficient duration to allow motor activity to approach steady-state levels during the last 20 percent of the session for control animals" (EPA 1998b, p. 39). There is no indication in the original study that that criterion was reached; indeed most motor-activity test sessions require at least 20 min to reach asymptotic levels (Fitzgerald et al. 1988). That deficiency of the study was not raised by EPA in its review. EPA also largely ignored the absence of exposure-response relationships for some behavioral end points.

No mode of action has been postulated for formaldehyde-induced neurologic effects. EPA concluded that behavioral changes seen in formaldehyde-exposed animals are unlikely to be attributable to the irritant properties of formaldehyde. The committee does not support that conclusion. For example, Sorg and Hochstatter (1999) observed alterations in formaldehyde-exposed rats in an odor-cued test of learning. It is possible that formaldehyde exposure resulted in olfactory epithelial injury sufficient to affect olfaction. Other studies (for example, Sorg et al. 2001) suggest that stress responses, such as altered cortisol concentrations, occur in formaldehyde-exposed animals. It is plausible that those changes occur because of nasal irritation and other local responses. Stress and related alterations in stress hormones are important potential confounders because they are associated with deficits in hippocampal-based memory function, alterations in hippocampal structure, and other neurologic responses (Pavlides et al. 2002; Conrad 2006; McEwen 2008; Zuena et al. 2008). Another concern raised by the committee is that the high reactivity of formaldehyde would not lend itself to substantial delivery to the nervous system.

The draft IRIS assessment indicates that there is some question as to whether formaldehyde should be considered a direct neurotoxicant (EPA 2010). Indeed, some portions of the assessment suggest that systemic effects are unexpected at formaldehyde concentrations less than 20 ppm. That idea is inconsistently presented in other parts of the document. The inconsistency in the document should be resolved.

## Hazard Identification and Use of EPA Guidelines

EPA has developed guidelines for neurotoxicity risk assessment (EPA 1998b). One cornerstone of the guidelines is the definition of neurotoxicity as an adverse change in the structure or function of the central or peripheral nervous system after exposure to an agent. Changes in motor activity, learning and memory, and other end points after formaldehyde exposure meet the definition of an adverse response. Although a mode of action for formaldehyde neurotoxicity is lacking, that gap does not preclude drawing a conclusion. The EPA guidelines state that "knowledge of exact mechanisms of action is not, however, necessary to conclude that a chemically induced change is a neurotoxic effect" (EPA 1998b, p. 10).

The neurotoxicity guidelines state that "the interpretation of data as indicative of a potential neurotoxic effect involves the evaluation of the validity of the database…There are four principal questions that should be addressed: whether the effects result from exposure (content validity); whether the effects are adverse or toxicologically significant (construct validity); whether there are correlative measures among behavioral, physiological, neurochemical, and morphological endpoints (concurrent validity); and whether the effects are predictive of what will happen under various conditions (predictive validity)" (EPA 1998b, p. 10). The draft IRIS assessment does not indicate whether those criteria were considered in the selection of the key studies. Indeed, data supporting concurrent and predictive validity are largely lacking for formaldehyde.

The EPA guidelines also state that "the minimum evidence necessary to judge that a potential hazard exists would be data demonstrating an adverse neurotoxic effect in a single appropriate, well-executed study in a single experimental animal species" (EPA 1998b, p. 53). There is concern that the selected studies are not sufficiently robust in design to be considered "well executed" for the purpose of neurotoxicity hazard identification. For example, motor-activity responses seen by Malek et al. (2003 a,b) in different test sessions in control animals were quite variable. Malek et al. also examined formaldehyde effects on learning and memory using a labyrinth swim maze, a test that could be affected by motor activity. Furthermore, the available human data have important shortcomings—such as limited exposure assessments and coexposures to neurotoxic solvents—that preclude a determination that formaldehyde is neurotoxic to humans.

## Study Selection for Calculation of Reference Concentration and Identification of Point of Departure

EPA concluded that the available epidemiologic studies did not provide sufficient exposure information to permit derivation of a point of departure for use in quantitative dose-response assessment. The draft IRIS assessment states that "confounding exposures to other neurotoxic solvents and inconsistent results prevent drawing definitive conclusions concerning the neurotoxicity of formaldehyde from these studies" (EPA 2010, p. 4-97). The committee agrees with EPA's decision not to use the human studies to calculate a candidate RfC.

Several studies in mice demonstrated dose-related neurotoxic effects after formaldehyde exposure. The studies were not considered for RfC development because the observed results might have been confounded by formaldehyde-induced reflex bradypnea and related physiologic responses. The committee agrees with EPA's decision not to use the experimental mouse studies for RfC development for the reasons cited.

EPA identified several experimental studies in rats that might be appropriate for candidate RfC development. According to EPA, the selected rat behavioral studies were not confounded by reflex bradypnea inasmuch as the effect occurs in rats only at doses above those at which the neurologic effects of concern were seen. EPA considered the studies by Malek et al. (2003a,c) that reported effects at low exposures to be the most robust. The draft IRIS assessment does not provide criteria that define why the studies were considered "robust" other than that the changes were observed at low concentrations. Malek et al. (2003c) found statistically significant reductions in motor activity after a single 2-hr exposure at 130-5,180 ppb (with testing 2 hr after cessation of exposure). Malek et al. (2003a) also showed a statistically significant reduction in performance on a learning task at similar exposures (100-5,400 ppb) when 2-hr exposures were repeated on 10 consecutive days ($p < 0.05$); performance was evaluated 2 hr after cessation of exposure, and concentration-related learning deficits were seen at all concentrations. The study was eventually selected as the key study by EPA. As noted earlier, no study was conducted according to existing EPA health-effects test guidelines for the conduct of a neurotoxicity screening battery or for evaluation of neurotoxicity end points (EPA 1998b). Accordingly, the studies have several methodologic shortcomings in how behavior was assessed by the investigators. In addition, neither study assessed subchronic or longer exposures; this draws into question their appropriateness for deriving a chronic RfC. The committee did not identify an alternative study that would be preferred for deriving a candidate RfC.

Malek et al. (2003a) reported a lowest observed-adverse-effect level (LOAEL) of 100 ppb in rats for neurologic and behavioral toxicity (impaired learning) after repeated exposure (2 hr/day over 10 days). A no-observed-adverse-effect level (NOAEL) was not identified for that effect. The committee notes that the point of departure for the study was subject to an exposure adjustment (Table 5-1 in the draft IRIS assessment). Testing was conducted 2 hr after exposure, and the duration was adjusted by EPA to 4 hr to include the entire period between start of exposure and testing. The committee disagrees with the duration adjustment because of the uncertainty in continuous-exposure adjustments for exposure durations as short as that used in the experimental study. The study was not carried forward for derivation of a candidate RfC, partly because of the uncertainty in extrapolating from the exposure conditions in the study to a chronic-exposure scenario; the committee agrees with EPA's decision in this regard.

## Conclusions and Recommendations

The committee concludes that the draft IRIS assessment overstates the evidence that formaldehyde is neurotoxic. The selected studies are not sufficiently robust in design to be considered well executed for the purpose of neurotoxicity-hazard identification. One study of rats by Malek et al. (2003a) was advanced by EPA for consideration. It was considered to offer information on an outcome relevant to humans at an appropriate concentration. Appropriately, the study was not used to calculate a candidate RfC, partly because of uncertainty in extrapolating from the exposure conditions in the study to a chronic-exposure scenario.

The committee recommends that EPA address the following in the revision of the formaldehyde draft IRIS assessment:

- Re-evaluate its conclusions that behavioral changes are unlikely to be related to irritant properties of formaldehyde.
- Resolve inconsistencies regarding the concentration at which systemic effects of formaldehyde exposure are expected. The draft IRIS assessment indicates that there is some question as to whether formaldehyde should be considered a direct neurotoxicant, and some portions of the assessment suggest that systemic effects are unexpected at formaldehyde concentrations less than 20 ppm. That statement is inconsistently made in other parts of the document.

## REPRODUCTIVE AND DEVELOPMENTAL TOXICITY

The reproductive and developmental outcomes considered in the draft IRIS assessment comprise a broad spectrum of specific outcomes, including infertility, low birthweight, spontaneous abortion, birth defects, functional deficits, and other altered health conditions. Each outcome may have a distinct pathogenesis and etiology. A variety of environmental, occupational, lifestyle, and genetic factors have been hypothesized to be associated with an increased risk of those outcomes.

Formaldehyde's potential mode of action for reproductive and developmental outcomes is uncertain; several modes have been suggested by animal studies, including endocrine disruption, genotoxic effects on gametes, and oxidative stress or damage. Critical questions remain about the association between inhalation exposure and the potential for adverse reproductive and developmental effects. EPA reviewed epidemiologic and animal studies that evaluated formaldehyde exposure in relation to fecundability (the per-cycle probability of conception), spontaneous abortion, birth defects, low birthweight, and reproductive effects (EPA 2010). EPA concluded that the epidemiologic studies provided evidence of a convincing relationship between occupational exposure to formaldehyde and adverse reproductive outcomes in women. EPA selected a single study (Taskinen et al. 1999) that evaluated the association between formaldehyde and fecundability, using time to pregnancy for determination of a candidate RfC.

## Study Identification

EPA appears to have conducted a thorough literature search and identified the available and appropriate studies. The epidemiologic literature is mixed in terms of study population, design, exposure conditions, and other factors. In addition to the epidemiologic studies, the draft IRIS assessment identifies 10 animal reproductive studies and 13 developmental studies (and one abstract) that reported reproductive and developmental effects after exposure to formaldehyde. Several animal studies were not described in their original publications in sufficient detail to assess their overall conclusions.

## Study Evaluation

The draft IRIS assessment generally follows EPA guidelines for the evaluation of reproductive toxicity (EPA 1996) and developmental toxicity (EPA 1991) regarding the consideration of key factors in the evaluation of epidemiologic studies in risk assessment, including study power, exposure measurement, selection bias, and confounding. Ideally, those factors would be reviewed and presented in an organized and systematic fashion for each study considered, the important biases would be systematically examined, and their potential magnitude and direction would be clearly enumerated. However, the descriptions and evaluation of the individual epidemiologic studies in the draft IRIS assessment are not consistent among studies. The characterization of the strengths and weaknesses of the studies varies; some studies receive a fuller treatment, including a more extensive assessment of bias and its consequences for estimating effect measures, and others receive less attention. For example, the evaluation of the study by Taskinen et al. (1994) of laboratory workers has minimal discussion of study strengths and weaknesses, and the discussion of bias is misleading or incomplete. Specifically, the discussion dismisses potential confounding by xylene exposure because the odds ratio for xylene and spontaneous abortion of 3.1 is less than the formalin odds ratio of 3.5 (EPA 2010, p. 4-88). The committee notes that although the odds ratio is slightly less for xylene, the estimates are the same for all practical purposes, and potential confounding by xylene exposure in the study remains an open question. The draft IRIS assessment also notes that exposure misclassification will not have "impacted the results of the study to any great extent" (EPA 2010, p. 4-88) but does not indicate the possible magnitude or direction of any exposure misclassification bias.

The committee disagrees with EPA's overall conclusion regarding the totality of the epidemiologic evidence related to the reproductive and developmental effects of formaldehyde. Specifically, the draft IRIS assessment states that "epidemiologic studies suggest a convincing relationship between occupational exposure to formaldehyde and adverse reproductive outcomes in women" (EPA 2010, p. 4-85). The committee, after assessing the literature, finds a suggestive pattern of association among a small number of studies rather than a convincing relationship. The committee's assessment is based on the overall pattern of positive association among most of the studies, but the generally limited exposure assessment and concern about other biases leads to the more appropriate descriptor of *suggestive* rather than *convincing*.

Many of the conclusions on developmental and reproductive effects in animal studies provided in Section 6.1.3.7 of the draft IRIS assessment are based on studies that are of questionable quality. The following statement seems to over interpret the results:

> Nevertheless, a number of animal studies have demonstrated effects of formaldehyde on pre- and postnatal development and on the reproductive system. For example, developmental toxicity was observed in two studies that evaluated a standard battery of developmental endpoints resulting from inhalation exposure on GDs [gestation days] 6-10 [Saillenfait et al. 1989; Martin 1990] (EPA 2010, p. 4-371).

Saillenfait et al. (1989) reported a statistically significant decrease in maternal weight gain at the highest concentration tested (40 ppm); this suggests maternal toxicity, but there were no statistically significant changes in number of implantations, resorptions, stage of resorptions, or live or dead fetuses. There was a statistically significant decrease in mean fetal bodyweights at 20 ppm in male fetuses with no associated maternal toxicity. Martin (1990) found no statistically significant differences in a standard battery of developmental end points, such as live or dead fetuses and implantation sites. There was decreased ossification of pubic bones in the 10-ppm (highest-concentration) group, but the group had a higher number of fetuses per litter associated with an overall lower weight. No other malformations were reported. Other studies were not described in sufficient detail to determine their quality. One study (Kilburn and Moro 1985) was published in abstract form only and probably should not be included in the analysis. In many of the studies, maternal toxicity, litter size, within-litter effects, and other quality-control measures were not discussed; this makes assessment of their quality difficult.

The draft IRIS assessment does not distinguish clearly between studies that are considered of high quality for use in risk assessment and studies that are considered for qualitative assessment only. With respect to the animal studies, it is not clear what weight was given to negative vs positive results. For example, in Section 4.4.9.1 of the draft, preimplantation loss in rats is discussed; two studies are given as supporting evidence (Sheveleva 1971; Kitaev et al. 1984), and eight inhalation studies are given as not reporting treatment-related embryolethality. A clear discussion that weighs both positive and negative results is needed.

Similarly, in the discussion of low birth weight and growth retardation (EPA 2010, Section 4.4.9.3), the effect of maternal toxicity is not discussed in the context of fetal growth retardation even though there is an indication of maternal toxicity in the effects summarized in Table 4-70. IARC (2006) considered 20 ppm to be a toxic concentration and reported that 10 ppm resulted in a significant decrease in food consumption. That statement suggests that maternal toxicity would occur after exposure at 10 ppm and greater and that effects noted at those doses should probably be attributed to maternal toxicity rather than to direct exposure to formaldehyde; this issue needs to be addressed in the evaluation of the studies. For example, maternal stress that can result from being put into inhalation chambers or from the irritating effects of formaldehyde at high concentrations is not discussed but could be a contributor to maternal toxicity. Because formaldehyde is a natural metabolic intermediate in humans and other animals, some discussion of the endogenous formaldehyde concentrations in the animal models is needed to put the exposures into context.

**Hazard Identification and Use of EPA Guidelines**

The draft IRIS assessment briefly summarizes the evaluation of key epidemiologic studies and touches on issues of bias (EPA 2010, Section 4.4.9). There is little discussion about the weight of evidence given to the studies. In a conclusion that there was sufficient evidence of fetal toxicity, some animal studies used as supporting evidence are of questionable quality. There is no discussion of whether formaldehyde (or its metabolites) could gain access to the fetus and cause adverse effects or whether the adverse effects were a result of maternal toxicity.

The review of animal studies in the draft IRIS assessment does not discuss the modes of action in sufficient detail to determine the biologic plausibility that formaldehyde adversely affects the fetus. Potential modes of action are endocrine disruption, genotoxic effects on gametes, and oxidative stress or damage (EPA 2010, Section 4.4.9.7). Although more weight is given to studies of animals exposed by inhalation, studies that used other exposure routes are included. A major concern in connection with developmental and reproductive toxicity is whether formaldehyde can penetrate past the portal of entry. That critical question affects the conclusions drawn from the animal studies, particularly those in which exposure was by a route other than inhalation. More emphasis is placed on studies whose route of exposure is inhalation because metabolism and distribution may differ substantially after oral, dermal, and intraperitoneal exposure. If formaldehyde does penetrate past the portal of entry, another important consideration is whether it crosses the placenta and gains entry into the fetus or whether it crosses the blood-testis barrier. There is little information in the draft IRIS assessment or in any of the reviewed studies concerning those points, and no attempts appear to have been made to measure formaldehyde or metabolites in target tissues.

Several conclusions are stated regarding the reproductive and developmental effects of inhalation exposure to formaldehyde (EPA 2010, Section 6.1.3.7). References are given to support the conclusions; however, the overall quality of the database is not discussed. For example, it is stated that "exposure of rat dams to formaldehyde during pregnancy has been shown to result in significantly decreased fetal weight gain" (EPA 2010, p. 6-13); but a statistically significant decrease in fetal weight, which was not associated with maternal toxicity, was reported in only one study in which male, but not female, fetuses were affected (Saillenfait et al. 1989).

Conclusions concerning male reproduction are similar; supporting studies generally used concentrations that result in significant weight loss and overt toxicity (Sarsilmaz et al. 1999; Ozen et al. 2002). The conclusions need to be placed into the context of potential confounders, such as maternal toxicity, stress from being placed in inhalation chambers, irritant concentrations above the odor threshold, and potential oral exposures by licking. The quality of the supporting studies needs to be stated clearly.

The overall database on developmental and reproductive effects of inhalation exposure to formaldehyde in animal studies is suggestive of an effect but not conclusive. When given by oral exposure or by injection, formaldehyde or its metabolites are capable of reaching reproductive tissues and the fetus. However, whether inhaled formaldehyde passes the portal of entry to access distant tissues—such as the gonads, hypothalamus, or the fetus—remains unresolved. In evaluating the animal data for reproductive effects, the draft IRIS assessment notes that there are no multigenerational tests for reproductive function (EPA 2010, Section 4.4.9.8). The committee agrees that that constitutes a data gap; particularly for male reproductive effects, such information is needed.

**Study Selection for Calculation of Reference Concentration and Identification of Point of Departure**

Regarding RfC derivation, EPA guidelines discuss methods to determine the adequacy of individual studies and the completeness of the overall database (EPA 2002). As noted above, the information on many health outcomes is not complete and does not contain critical details, such as duration and timing of exposure. Furthermore, in reviewing critical studies, the draft IRIS assessment does not adequately address important aspects of its guidance, such as the following:

- Is there sufficient description of the protocol, statistical analyses, and results to make an evaluation?
- Were appropriate statistical techniques applied for each end point, and was the power of the study adequate to detect effects?
- Did the study establish dose-response relationships?
- Are the results of the study biologically plausible?

The epidemiologic studies provide only a suggestive pattern of association. However, to be consistent with EPA guidelines, an RfC can be calculated by using the best available study evidence. EPA chose the study by Taskinen et al. (1999) to derive a candidate RfC. That study of female Finnish wood workers examined estimated workplace formaldehyde exposure primarily in relation to time to pregnancy but secondarily to other outcomes, including endometriosis and spontaneous abortion. The study had multiple strengths, including the national identification of workers and birth outcomes; industrial-hygienist assessment of potential exposures, including workplace measurements; and adjustment for multiple potential confounders, including other exposures. The study weaknesses included the use of a mailed questionnaire for exposure and covariate information; potential recall bias as to work tasks; no consideration of work accidents; inadequate description of exposure sources, such as the number of measurements taken; and the use of measurements from workplaces other than their own specific workplace that varied by exposure category. Furthermore, participant response to the questionnaire was less than outstanding.

EPA chose that study from the available epidemiologic studies of reproductive effects because of its overall strengths, the low likelihood of an important effect of selection bias, and consistency with other epidemiologic studies and animal evidence on fetal loss (EPA 2010). Furthermore, the draft IRIS assessment notes that the study population was well defined and adequately selected to allow examination of health effects in people who had different exposures. The committee agrees that it has a number of important strengths compared with the other reproductive epidemiologic studies evaluated. Notable strengths include exposure assessment, a relatively easily measured outcome (time to pregnancy), and assessment of confounding, such as by occupational exposures.

The draft IRIS assessment indicates that the study could be used for three outcomes: miscarriage, endometriosis, and decreased fecundity density ratio (FDR). However, because of the concerns about the miscarriage and endometriosis analyses, the FDR results were chosen as the critical effect for a candidate RfC. For example, the spontaneous-abortion analysis was not the primary aim, and the exposure and response pattern was not consistent with the increased risk found in the low-exposure group. In addition, the spontaneous-abortion analysis did not adjust for all covariates used in the FDR analysis. EPA also noted that the endometriosis results may be confounded by other solvents. The committee agrees that the choice of outcome from Taskinen et al. (1999) is appropriate for the reasons provided in the draft IRIS assessment.

However, the committee is concerned that basing an RfC on a single human study in a minimal human database is problematic. EPA guidelines state that "a reference value based on a single study would likely have a high degree of uncertainty" (EPA 2002, p. 4-20). Although multiple studies of varied quality have assessed spontaneous abortions, the study by Taskinen et al. (1999) is the only one that measured time to pregnancy.

## Conclusions and Recommendations

The review of the reproductive and developmental outcomes in the draft IRIS assessment includes relevant outcomes and literature. It does not consistently provide a critical evaluation of the quality of publications and data presented or note strengths and weaknesses of each study. That is especially the case with the animal studies. The rationale for the assessment of the body of the epidemiologic evidence as convincing is not well articulated. Issues regarding the potential portal of entry and mode of action in relation to reproductive and developmental outcomes are not integrated into the weight-of-evidence discussion. Nonetheless, despite the shortcomings in the database and aspects of the review, the most relevant epidemiologic study and specific outcome are advanced for derivation of a candidate RfC. The point of departure is appropriately selected.

The committee recommends that EPA address the following in the revision of the formaldehyde draft IRIS assessment:

- Provide a consistent critical evaluation of the study quality and data presented, particularly strengths and weaknesses of each study. That is especially needed for the animal studies.
- Articulate better the basis of the assessment of the epidemiologic evidence.
- Integrate better the issues surrounding systemic delivery and mode of action for reproductive and developmental outcomes into the weight-of-evidence discussion.

## LYMPHOHEMATOPOIETIC CANCERS

Lymphohematopoietic (LHP) cancers are a heterogeneous group of cancers that encompass a wide variety of leukemias and lymphomas. Although they all arise from the hematopoietic system, these cancers are often derived from cells of different origin, can demonstrate unique genetic abnormalities, and may arise in different tissues (Figure 5-1). Those differences indicate that their etiologic bases may be distinct.

Although the draft IRIS assessment explores specific diagnoses—such as acute myeloid leukemia (AML), chronic myeloid leukemia (CML), and Hodgkin lymphoma and multiple myeloma (see, for example, EPA 2010, Table 4-92)—the determinations of causality are made for the heterogeneous groupings "all LHP cancers," "all leukemias," and "myeloid leukemias." The grouping "all LHP cancers" includes at least 14 biologically distinct diagnoses in humans (Figure 5-1) and should not be used in determinations of causality. The draft IRIS assessment should include information about the relative incidence of the

# Systemic Health Effects

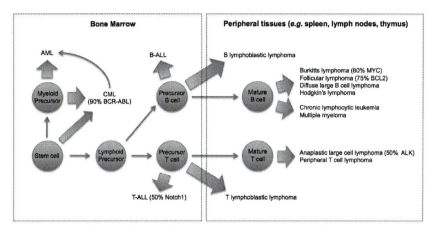

**FIGURE 5-1** Origins of lymphohematopoietic cancers. Cells of origin, common genetic abnormalities, and tissues of origin are indicated for diverse hematopoietic malignancies. Compared with Figure 4-32 in the draft IRIS assessment, this figure clarifies the distinct cells of origin of acute myeloblastic leukemia (AML), T-cell acute lymphoblastic leukemia (T-ALL), B-cell acute lymphoblastic leukemia (B-ALL), and most mature leukemias and lymphomas. Abbreviations: ALK, anaplastic lymphoma kinase; BCL2, B-cell leukemia 2; BCR-ABL, breakpoint cluster region-Abelson murine leukemia; and MYC, myelocytomatosis viral oncogene homolog.

leukemia subtypes because the contribution of AML, CML, "myeloid leukemias," and "all leukemias" to the grouping "all LHP" may not be obvious to readers and may help with interpretation (Figure 5-2).

Another important topic of discussion in the draft IRIS assessment is that of potential modes of action of formaldehyde as a cause of diverse LHP cancers. As discussed in Chapter 3 of the present report, the available experimental data indicate that formaldehyde itself does not penetrate beyond the superficial layer of the portal of entry, the epithelium of the nasopharynx. Therefore, long-used models of chemical leukemogenesis in which there is direct toxicity to hematopoietic cells in the bone marrow are unlikely to explain the proposed distal effect of formaldehyde on hematopoietic precursors. However, evidence of formaldehyde-induced DNA adducts and DNA damage in circulating lymphocytes suggests that hematopoietic cells might be affected by inhaled formaldehyde, presumably at the nasal epithelium or nasal-associated lymphoid tissue (NALT). EPA and others propose a model in which lymphoid precursors or hematopoietic stem cells circulate or migrate to the nasal epithelium, where they are directly exposed to formaldehyde and ultimately result in diverse LHP cancers. As the draft IRIS assessment states, that hypothesis seems plausible for Hodgkin lymphoma and multiple myeloma, which arise from precursors in the peripheral

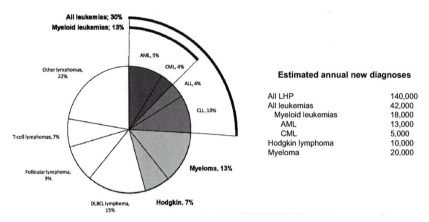

**FIGURE 5-2** Relative incidence and estimated annual new diagnoses of common lymphohematopoietic cancer subtypes in the United States. Abbreviations: ALL, acute lymphoblastic leukemia; AML, acute myeloid leukemia; CLL, chronic lymphoblastic leukemia; CML, chronic myeloid leukemia; DLBCL, diffuse large B-cell lymphoma; and LHP, lymphohematopoietic. Figure based on data from ACS (2010), LLS (2011), and SEER (2010).

tissues. However, inasmuch as experimental evidence is absent, there is no direct support for the hypothesis. Similarly, although recent evidence demonstrates that normal hematopoietic precursors do indeed leave the bone marrow and circulate as part of a daily circadian rhythm (Mendez-Ferrer et al. 2009), studies have not shown that the cells are present in the nasal epithelium or NALT, nor have they shown formaldehyde-induced effects in vivo. An additional hypothesis is that formaldehyde exposure at the port of entry induces secondary systemic effects, such as immune modulation or systemic inflammation, both of which are associated with LHP cancers. However, given the lack of direct data that could support those hypothetical modes of action, EPA could shorten those sections of the draft substantially and note that the modes of action remain uncertain.

Similarly, there is a paucity of evidence of formaldehyde-induced LHP cancers in animal models. EPA's unpublished re-analysis of the Battelle chronic inhaled formaldehyde experiments in mice and rats (Battelle Columbus Laboratories 1981), although intriguing, provides the only positive findings and thus does not contribute to the weight of evidence of causality.

## Study Identification

The draft IRIS assessment comprehensively presents studies available through late 2009 that evaluate formaldehyde exposure and risk of LHP cancers. The draft provides commentary on multiple studies that had negative and positive findings, cohorts that were the subject of multiple analyses or publications, and meta-analyses. The emphasis on studies of occupational cohorts is appropriate, given that they provide the most specific and detailed exposure assessment that can be applied in risk assessments. The committee is not aware of any important studies that are missing from the analysis, although several relevant studies have been published since the draft was released (for example, Andersen et al. 2010; Bachand et al. 2010; Lu et al. 2010; Schwilk et al. 2010). To make the IRIS assessment as timely as possible, inclusion of the recent studies seems warranted in the revision.

## Study Evaluation

EPA's review is extensive and covers in considerable detail a substantial body of pertinent epidemiologic and toxicologic literature. The study evaluations tend to be long narratives that provide substantial detail on reported findings. The heterogeneity of LHP cancers is acknowledged as a complicating factor in assessing causation and ultimately in the cancer assessment.

However, there is no clearly articulated framework for establishing causation on the basis of the weight and strength of evidence. An a priori presentation of the study selection criteria (for example, quality of exposure assessment, control of confounding variables, and statistical power) is also missing. Both the framework and study selection criteria are critical for any determination of causation. For example, the concept of consistency of findings within and among studies is not defined, so it is difficult to determine how studies with different study populations, cancer incidence, and exposure measures were combined to provide a consistent and conclusive determination of causality. As a result, the conclusion of causation appears to be based on a subjective view of the overall data. Given the limitations of the epidemiologic studies (particularly uncertainties of exposure assessment, possible confounding by other pollutants, and reliance on mortality rather than incidence data), a clear statement and consistent use of the weight-of-evidence criteria would strengthen the conclusions. Additional explicit reference to the EPA *Guidelines for Carcinogen Risk Assessment* (EPA 2005a) and the *Supplemental Guidance for Assessing Susceptibility from Early-Life Exposure to Carcinogens* (EPA 2005b) or other guidelines or precedent for assessing the quality and importance of studies is recommended.

The absence of a causation framework is especially problematic for the individual LHP cancers, given the highly variable epidemiologic literature and the high uncertainty of mode of action. Important differences exist in the reported findings of the most influential studies—of UK (Coggon et al. 2003) and U.S. (Beane-Freeman et al. 2009) industrial cohorts, the National Institute for Occupational Safety and Health garment workers cohort (Pinkerton et al. 2004), and U.S. embalmers (Hauptmann et al. 2004). The differences should be discussed and weighed, specifically as to how they were taken into account in EPA's determinations of causality. For example, in the highly influential National Cancer Institute (NCI) cohort study (Beane-Freeman et al. 2009), the strength and specificity of the exposure-response associations varied considerably over the period in which the cohort was followed. In addition, the reliance on the peak-exposure metric to determine causality in that study rather than the more conventional dose metric of cumulative exposure should be further justified, particularly in the absence of established modes of action.

### Hazard Identification and Use of EPA Guidelines

The hazard identification concluded that there is a causal association between formaldehyde exposure and mortality from all LHP cancers, all leukemias as a group, myeloid leukemias, Hodgkin lymphoma, and multiple myeloma. As noted earlier, the committee strongly discourages the use of the grouping "all LHP cancers," given the biologic heterogeneity within this group. For the other groupings, the committee finds that the conclusions of causality are not adequately supported by the current narrative. Further discussion of LHP subtype diagnosis in various studies would aid in comparison of findings. Sections covering all leukemias and myeloid leukemias are adequate in depth but would benefit from a clearer synthesis of the data and more explicit reference to guidelines or precedents for evaluating evidence when there are notable differences in data quality and conflicting results among studies. The committee recommends caution on EPA's part in using meta-analyses performed by others to assess causality or to quantify effects. Meta-analysis can be a valuable method for summarizing evidence but can also be subject to variable interpretations depending on how literature is selected and reviewed and data analyzed. Given the conflicting conclusions of published meta-analyses of formaldehyde and LHP cancers (see, for example, Zhang et al. 2009; Bachand et al. 2010; Schwilk et al. 2010), EPA is encouraged to perform its own meta-analysis *if* the agency chooses to use meta-analysis as a tool to assess causation.

### Study Selection for Calculation of Unit Risk

As articulated by EPA, few studies can be used to calculate risk estimates. Regardless, the selection and use of the NCI cohort (Beane-Freeman et al. 2009) should be further justified. Indeed, interpretation of the study results is not straightforward given that the findings differ from those of earlier analyses of the same cohort and differ for peak, average, and cumulative formaldehyde exposures. In the absence of evidence regarding exposure-disease mechanisms, as in the case of formaldehyde and LHP cancers, cumulative exposure is typically the default dose metric applied in epidemiologic analyses and risk assessment. But the most significant results were found for peak exposures, which have the greatest associated uncertainty. In view of the importance of this study, EPA should clarify the basis of its interpretations of the results regarding the various

*Systemic Health Effects* 113

dose metrics and the various LHP cancers. Despite those concerns, the committee agrees that the NCI study is the most appropriate available to carry forward for calculation of the unit risk.

## Conclusions and Recommendations

The committee recommends that EPA address the following in the revision of the formaldehyde draft IRIS assessment:

- Focus on the most specific diagnoses available in the epidemiologic data, such as acute myeloblastic leukemia (International Classification of Diseases [ICD] 205.0), chronic lymphocytic leukemia (ICD 204.1), and specific lymphomas, such as Burkitt (ICD 200.2), Hodgkin (ICD 201), anaplastic large-cell (ICD 200.6), and peripheral T-cell lymphoma (ICD 202.7). The committee does not support consideration of the grouping "all LHP cancers" because this grouping combines diverse cancers that are not closely related in cells of origin and in other characteristics.
- Evaluate existing studies and data with concise discussions of background and speculative hypotheses. The narratives in the draft IRIS assessments are sometimes too long and unfocused.
- Clarify how EPA determined weight and strength of evidence. The draft assessment should be revised to discuss the benefits, limitations, and justifications of using one exposure metric to determine causality and another to calculate cancer unit risk. Because the draft assessment relies solely on epidemiologic studies to determine causality, further discussion of the specific strengths, weaknesses, and inconsistencies in several key studies is needed. As stated in EPA's cancer guidelines, EPA's approach to weight of evidence should include "a single integrative step after assessing all of the individual lines of evidence" (EPA 2005a, Section 1.3.3, p. 1-11). Although a synthesis and summary are provided, the process that EPA used to weigh different lines of evidence and how that evidence was integrated into a final conclusion are not apparent in the draft assessment and should be made clear in the final version.
- Revisit arguments that support determinations of causality of specific LHP cancers and in so doing include detailed descriptions of the criteria that were used to weigh evidence and assess causality. That will add needed transparency and validity to the conclusions.
- If EPA decides to rely on meta-analysis as a tool to assess causation, it should perform its own meta-analysis with particular attention to specific diagnoses and to variables selected and combined for analysis. The contrasting conclusions of the published meta-analyses make it difficult to rely on conclusions from any one analysis (see, for example, Zhang et al. 2009; Bachand et al. 2010; Schwilk et al. 2010).

## REFERENCES

ACS (American Cancer Society). 2010. Cancer Facts & Figures 2010. American Cancer Society [online]. Available: http://www.cancer.org/acs/groups/content/@epide miologysurveilance/documents/document/acspc-026238.pdf [accessed March 2, 2011].

Adams, D.O., T.A. Hamilton, L.D. Lauer, and J.H. Dean. 1987. The effect of formaldehyde exposure upon the mononuclear phagocyte system of mice. Toxicol. Appl. Pharmacol. 88(2):165-174.

Andersen, M.E., H.J. Clewell III, E. Bermudez, D.E. Dodd, G.A. Willson, J.L. Campbell, and R.S. Thomas. 2010. Formaldehyde: Integrating dosimetry, cytotoxicity, and genomics to understand dose-dependent transitions for an endogenous compound. Toxicol. Sci. 118(2):716-731.

Bachand, A.M., K.A. Mundt, D.J. Mundt, and R.R. Montgomery. 2010. Epidemiological studies of formaldehyde exposure and risk of leukemia and nasophryngeal cancer: A meta-analysis. Crit. Rev. Toxicol. 40(2):85-100.

Battelle Columbus Laboratories. 1981. Final Report on a Chronic Inhalation Toxicology Study in Rats and Mice Exposed to Formaldehyde. Prepared by Battelle Columbus Laboratories, Columbus, OH, for the Chemical Industry Institute of Toxicology (CIIT), Research Triangle Park, NC. CIIT Docket No. 10922.

Beane-Freeman, L.E., A. Blair, J.H. Lubin, P.A. Stewart, R.B. Hayes, R.N. Hoover, and M. Hauptmann. 2009. Mortality from Lymphohematopoietic malignancies among workers in formaldehyde industries: The National Cancer Institute cohort. J. Natl. Cancer Inst. 101(10):751-761.

Coggon, D., E.C. Harris, J. Poole, and K.T. Palmer. 2003. Extended follow-up of a cohort of British chemical workers exposed to formaldehyde. J. Natl. Cancer Inst. 95(21):1608-1615.Conrad, C.D. 2006. What is the functional significance of chronic stress-induced CA3 dendritic retraction within the hippocampus? Behav. Cogn. Neurosci. 5(1):41-60.

Dean, J.H., L.D. Lauer, R.V. House, M.J. Murray, W.S. Stillman, R.D. Irons, W.H. Steinhagen, M.C. Phelps, and D.O. Adams. 1984. Studies of immune function and host resistance in B6C3F1 mice exposed to formaldehyde. Toxicol. Appl. Pharmacol. 72(3):519-529.

EPA (U.S. Environmental Protection Agency). 1991. Guidelines for Developmental Toxicity Risk Assessment. EPA/600/FR-91/001. Risk Assessment Forum, U.S. Environmental Protection Agency, Washington, DC. December 1991 [online]. Available: http://cfpub.epa.gov/ncea/cfm/recordisplay.cfm?deid=23162 [accessed Nov. 24, 2010].

EPA (U.S. Environmental Protection Agency). 1996. Guidelines for Reproductive Toxicity Risk Assessment. EPA/630/R-96/009. Risk Assessment Forum, U.S. Environmental Protection Agency, Washington, DC. October 1996 [online]. Available: http://www.epa.gov/raf/publications/pdfs/REPRO51.PDF [accessed Nov. 24, 2010].

EPA (U.S. Environmental Protection Agency). 1998a. Health Effects Test Guidelines OPPTS 870.7800 Immunotoxicity. U.S. Environmental Protection Agency [online]. Available: http://www.regulations.gov/#!documentDetail;D=EPA-HQ-OPPT-2009-0156-0049 [accessed Feb. 22, 2011].

EPA (U.S. Environmental Protection Agency). 1998b. Guidelines for Neurotoxicity Risk Assessment. EPA/630-95/001F. Risk Assessment Forum, U.S. Environmental

Protection Agency, Washington, DC. April 1998 [online]. Available: http://www.epa.gov/raf/publications/pdfs/NEUROTOX.PDF [accessed Nov. 24, 2010].

EPA (U.S. Environmental Protection Agency). 2002. A Review of the Reference Dose and Reference Concentration Processes. External Review Draft. EPA/630/P-02/002A. Reference Dose/Reference Concentration (RfD/RfC) Technical Panel, Risk Assessment Forum, U.S. Environmental Protection Agency, Washington, DC [online]. Available: http://www.epa.gov/raf/publications/pdfs/rfdr fcextrevdrft.pdf [accessed Jan. 6, 2010].

EPA (U.S. Environmental Protection Agency). 2005a. Guidelines for Carcinogen Risk Assessment. EPA/630/P-03/001F. Risk Assessment Forum, U.S. Environmental Protection Agency, Washington, DC. March 2005 [online]. Available: http://www.epa.gov/raf/publications/pdfs/CANCER_GUIDELINES_FINAL_3-25-05.PDF [accessed Nov. 24, 2010].

EPA (U.S. Environmental Protection Agency). 2005b. Supplemental Guidance for Assessing Susceptibility from Early-Life Exposure to Carcinogens. EPA/630/R-03/003F. Risk Assessment Forum, U.S. Environmental Protection Agency, Washington, DC. March 2005 [online]. Available: http://www.epa.gov/ttn/at w/childrens_supplement_final.pdf [accessed Nov. 24, 2010].

EPA (U.S. Environmental Protection Agency). 2010. Toxicological Review of Formaldehyde (CAS No. 50-00-0) – Inhalation Assessment: In Support of Summary Information on the Integrated Risk Information System (IRIS). External Review Draft. EPA/635/R-10/002A. U.S. Environmental Protection Agency, Washington, DC [online]. Available: http://cfpub.epa.gov/ncea/iris_drafts/rec ordisplay.cfm?deid=223614 [accessed Nov. 22, 2010].

Fitzgerald, R.E., M. Berres, and U. Schaeppi. 1988. Validation of a photobeam system for assessment of motor activity in rats. Toxicology 49(2-3):433-439.

Hauptmann, M., J.H. Lubin, P.A. Stewart, R.B. Hayes, and A. Blair. 2004. Mortality from solid cancers among workers in formaldehyde industries. Am. J. Epidemiol. 159(12):1117-1130.

Holmström, M., and B. Wilhelmsson. 1988. Respiratory symptoms and pathophysiological effects of occupational exposure to formaldehyde and wood dust. Scand. J. Work Environ. Health 14(5):306-311.

Holness, D.L., and J.R. Nethercott. 1989. Health status of funeral service workers exposed to formaldehyde. Arch. Environ. Health 44(4):222-228.

IARC (International Agency for Research on Cancer). 2006. Formaldehyde. Pp. 39-325 in Formaldehyde, 2-Butoxyethanol and 1-tert-Butoxypropan-2-ol. IARC Monographs on the Evaluation of Carcinogenic Risks to Humans, Vol. 88. Lyon, France: World Health Organization, International Agency for Research on Cancer.

Kilburn, K.H. 1994. Neurobehavioral impairment and seizures from formaldehyde. Arch. Environ. Health 49(1):37-44.

Kilburn, K.H., and A. Moro. 1985. Reproductive and maternal effects of formaldehyde (HCHO) in rats. Fed. Proc. 44:535.

Kilburn, K.H., B.C. Seidman, and R. Warshaw. 1985. Neurobehavioral and respiratory symptoms of formaldehyde and xylene exposure in histology technicians. Arch. Environ. Health 40(4):229-233.

Kilburn, K.H., R. Warshaw, and J.C. Thornton. 1987. Formaldehyde impairs memory equilibrium and dexterity in histology technicians: Effects which persist for days after exposure. Arch. Environ. Health 42(2):117-120.

Krzyzanowski, M., J.J. Quackenboss, and M.D. Lebowitz. 1990. Chronic respiratory effects of indoor formaldehyde exposure. Environ. Res. 52(2):117-125.

Lino dos Santos Franco, A., H.V. Domingos, A.S. Damazo, A.C. Breithaupt-Faloppa, A.P. de Oliviera, S.K. Costa, S.M. Oliani, R.M. Oliveira-Filho, B.B. Vargaftig, and W. Tavares-de-Lima. 2009. Reduced allergic lung inflammation in rats following formaldehyde exposure: Long term effects on multiple effector systems. Toxicology 256(3):157-163.

LLS (The Leukemia & Lymphoma Society). 2011. Facts 2010-2011. The Leukemia & Lymphoma Society [online]. Available: http://www.lls.org/content/nationalco ntent/resourcecenter/freeeducationmaterials/generalcancer/pdf/facts [accessed March 2, 2011].

Lu, K., L.B. Collins, H. Ru, E. Bermudez, and J.A. Swenburg. 2010. Distribution of DNA adducts caused by inhaled formaldehyde is consistent with induction of nasal carcinoma but not leukemia. Toxicol. Sci. 116(2):441-451.

Lyapina, M., G. Zhelezova, E. Petrova, and M. Boev. 2004. Flow cytometric determination of neutrophil burst activity in workers exposed to formaldehyde. Int. Arch. Occup. Environ. Health 77(5):335-340.

Malek, F.A., K.U. Möritz, and J. Fanghänel. 2003a. A study on the effect of inhalative formaldehyde exposure on water labyrinth test performance in rats. Ann. Anat. 185(3):277-285.

Malek, F.A., K.U. Möritz, and J. Fanghänel. 2003b. Formaldehyde inhalation and open field behaviour in rats. Indian J. Med. Res. 118:90-96.

Malek, F.A., K.U. Möritz, and J. Fanghänel. 2003c. A study on specific behavioral effects of formaldehyde in the rat. J. Exp. Anim. Sci. 42(3):160-170.

Martin, W.J. 1990. A teratology study of inhaled formaldehyde in the rat. Reprod. Toxicol. 4(3):237-239.

McEwen, B.S. 2008. Central effects of stress hormones in health and disease: Understanding the protective and damaging effects of stress and stress mediators. Eur. J. Pharmacol. 583(2-3):174-185.

Mendez-Ferrer, S., A. Chow, M. Merad, and P.S. Frenette. 2009. Circadian rhythms influence hematopoietic stem cells. Curr. Opin. Hematol. 16(4):235-242.

Morgan, K.T., D.L. Patterson, and E.A. Gross. 1986. Responses of the nasal mucociliary apparatus of F-344 rats to formaldehyde gas. Toxicol. Appl. Pharmacol. 82(1):1-13.

Ohtsuka, R., Y. Shutoh, H. Fujie, S. Yamahuchi, M. Takeda, T. Harada, and K. Doi. 2003. Rat strain difference in histology and expression of Th1- and Th2- related cytokines in nasal mucosa after short-term formaldehyde inhalation. Exp. Toxicol. Pathol. 54(4):287-291.

Ozen, O.A., M. Yaman, M. Sarsilmaz, A. Songur, and I. Kus. 2002. Testicular zinc, copper and iron concentrations in male rats exposed to subacute and subchronic formaldehyde gas inhalation. J. Trace Elem. Med. Biol. 16(2):119-122.

Pavlides, C., L.G. Nivón, and B.S. McEwen. 2002. Effects of chronic stress on hippocampal long-term potentiation. Hippocampus 12(2):245-257.

Pinkerton, L.E., M.J. Hein, and L.T. Stayner. 2004. Mortality among a cohort of garment workers exposed to formaldehyde: An update. Occup. Environ. Med. 61(3):193-200.

Riedel, F., E. Hasenauer, P.J. Barth, A. Koziorowski, and C.H. Rieger. 1996. Formaldehyde exposure enhances inhalative allergic sensitization in the guinea pig. Allergy 51(2):94-99.

Sadakane, K., H. Takano, T. Ichinose, R. Yanagisawa, and T. Shibamoto. 2002. Formaldehyde enhances mite allergen-induced eosinophilic inflammation in the murine airway. J. Environ. Pathol. Toxicol. Oncol. 21(3):267-276.

Saillenfait, A.M., P. Bonnet, and J. de Ceaurriz. 1989. The effects of maternally inhaled formaldehyde on embryonal and foetal development in rats. Food Chem. Toxicol. 27(8):545-548.

Sarsilmaz, M., O.A. Ozen, N. Akpolat, I. Kus, and A. Songur. 1999. The histopathologic effects of inhaled formaldehyde on Leydig cells of the rats in subacute period [in Turkish]. J. Med. (Firat University Health Science) 13(1):37-40.

Schwilk, E., L. Zhang, M.T. Smith, A.H. Smith, and C. Steinmaus. 2010. Formaldehyde and leukemia: An updated meta-analysis and evaluation of bias. J. Occup. Environ. Med. 52(9):878-886.

SEER (Surveillance Epidemiology and End Results). 2010. Estimated New Cancer Cases and Deaths for 2010: All Races, By Sex. American Cancer Society, Surveillance Epidemiology and End Results [online]. Available: http://seer.cancer.gov/csr/1975_2007/results_single/sect_01_table.01.pdf [accessed March 2, 2011].

Sorg, B.A., and T. Hochstatter. 1999. Behavioral sensitization after repeated formaldehyde exposure in rats. Toxicol. Ind. Health 15(3-4):346-355.

Sorg, B.A., T.M. Bailie, M.L. Tschirgi, N. Li, and W.R. Wu. 2001. Exposure to repeated low-level formaldehyde in rats increases basal corticosterone levels and enhances the corticosterone response to subsequent formaldehyde. Brain Res. 898(2):314-320.

Tarkowski, M., and P. Gorski. 1995. Increased IgE antiovalbumin level in mice exposed to formaldehyde. Int. Arch. Allergy Immunol. 106(4):422-424.

Taskinen, H., P. Kyyronen, K. Hemminki, M. Hoikkala, K. Lajunen, and M.L. Lindbohm. 1994. Laboratory work and pregnancy outcome. J. Occup. Med. 36(3):311-319.

Taskinen, H.K., P. Kyyronen, M. Sallmen, S.V. Virtanen, T.A. Liukkonen, O. Huida, M.L. Lindbohm, and A. Anttila. 1999. Reduced fertility among female wood workers exposed to formaldehyde. Am. J. Ind. Med. 36(1):206-212.

Weisskopf, M., N. Morozova, E.J. O'Reilly, M.L. McCullough, E.E. Calle, M.J. Thun, and A. Ascherio. 2009. Prospective study of chemical exposures and amyotrophic lateral sclerosis mortality. J. Neurol. Neurosurg. Psychiatry 80(5):558-561.

Zuena, A.R., J. Mairesse, P. Casolini, C. Cinque, G.S. Alemà, S. Morley-Fletcher, V. Chiodi, L.G. Spagnoli, R. Gradini, A. Catalani, F. Nicoletti, and S. Maccari. 2008. Prenatal restraint stress generates two distinct behavioral and neurochemical profiles in male and female rats. PLoS One 3(5):e2170.

Zhang, L., C. Steinmaus, D.A. Eastmond, X.K. Xin, and M.T. Smith. 2009. Formaldehyde exposure and leukemia: A new meta-analysis and potential mechanisms. Mutat. Res. 681(2-3):150-168.

# 6

# Reference Concentrations for Noncancer Effects and Unit Risks for Cancers

Chapter 5 of the draft IRIS assessment discusses the derivation of reference concentrations (RfCs) for noncancer effects and unit risks for cancers. Because estimates of RfCs and unit risks are subject to uncertainty and variation at every stage of the computational process, the committee conducted a thorough appraisal of the Environmental Protection Agency (EPA) process and analysis for calculating the estimates. In this chapter, the committee provides its review of EPA's derivation of RfCs and unit risks and offers its conclusions and recommendations regarding these two key products of the IRIS assessment.

The committee notes that EPA's dose-response assessments for cancer and noncancer effects have evaluated some end points for which there may not be adequate evidence to support the conclusion of a causal relationship between that end point and formaldehyde exposure. For example, modes of action for leukemia and Hodgkin lymphoma remain questionable, as noted by the present committee at various places in this report (Chapters 3 and 5). The committee recognizes, however, that EPA has followed its various risk-assessment guidelines (EPA1991, 1998, 2005) in conducting the dose-response assessments. In cancer risk assessment, for example, "dose-response assessments are generally completed for agents considered 'carcinogenic to humans' and 'likely to be carcinogenic to humans'" (EPA 2005). Dose-response assessments include an analysis of all tumor types on the basis of potential causality of the agent and may be conducted to provide a sense of the magnitude and uncertainty of potential risks, especially when the evidence is provided from a well-conducted study (EPA 2005). It is within that framework that the present committee reviewed EPA's calculation of RfCs for noncancer effects and unit risks for cancer. The review is partly geared toward an analysis of uncertainties associated with the underlying risk estimates and is not necessarily an endorsement, for example, of using a specific cancer, such as leukemia, for a consensus risk estimate. The committee's opinions on mode of action and weight of evidence concerning specific health outcomes are given in Chapters 3-5 of the present report.

## FORMALDEHYDE REFERENCE CONCENTRATIONS

EPA defines an RfC as "an estimate (with uncertainty spanning perhaps an order of magnitude) of a continuous inhalation exposure to the human population (including sensitive subgroups) that is likely to be without an appreciable risk of deleterious effects during a lifetime" (EPA 2010a). That is, an RfC is a concentration at which exposures would be allowed to occur with sufficient certainty, taking into account susceptibility and variability, that adverse outcomes would not result. RfCs are used by EPA, state agencies, various regulatory agencies, and other entities to develop allowable ambient air concentrations and to evaluate risks posed by current and potential exposures.

The draft IRIS assessment proposes several RfCs for formaldehyde that are based on "three studies of related health effects: asthma, allergic sensitization, pulmonary function, and symptoms of respiratory disease in children from in-home exposure to formaldehyde" (Rumchev et al. 2002; Garrett et al. 1999; Krzyzanowski et al. 1990) (EPA 2010b, p. 5-66). The discussion concludes by presenting a range (1-9 ppb), rather than a specific value, for the RfC. The committee was asked to comment on values of the uncertainty factors used to derive the RfCs that account for human population variability and for deficiencies in the overall database (see Box 1-1).

Chapters 4 and 5 of the present report addressed the health effects associated with formaldehyde exposure and reviewed the candidate critical effects, relevant studies, and points of departure identified by EPA. EPA's process for developing the RfC for formaldehyde is illustrated in Figure 6-1. The following sections briefly summarize EPA's selection of critical effects and key studies and identification of points of departure for derivation of candidate RfCs. Information that is relevant to evaluating the uncertainty factors proposed by EPA is then presented, and the committee provides its recommendations for those factors. Finally, the committee comments on the IRIS process for derivation of RfCs and provides suggestions for improving the process of selecting a final RfC.

### Selection of Candidate Noncancer Effects

Health effects associated with formaldehyde exposure have been studied extensively in people, laboratory animals, and in vitro systems with a variety of study designs. EPA evaluated a broad array of health effects that the committee

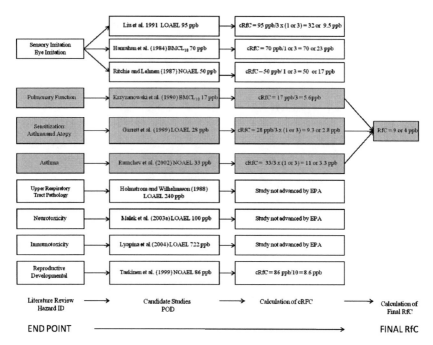

**FIGURE 6-1** Illustration of EPA's process for deriving a reference concentration for formaldehyde. Abbreviations: ID, identification; RfC, reference concentration; cRFC, candidate reference concentration; POD, point of departure; EPA, Environmental Protection Agency; LOAEL, lowest observed adverse effect-level; NOAEL, no observed adverse effect level; and $BMCL_{10}$, lower 95% confidence limit on the benchmark concentration corresponding to a 10% response level.

characterized as portal-of-entry or systemic. For portal-of-entry noncancer effects, the draft IRIS assessment concludes that formaldehyde causes sensory irritation, decreased pulmonary function, histopathologic lesions of the upper respiratory tract, and asthma and allergic sensitization. The committee agrees with EPA's assessment of a causal relationship between formaldehyde and those effects with the exception of incident asthma, which was based on the study by Rumchev et al. (2002). As noted in Chapter 4, the draft IRIS assessment does not sufficiently consider the complexities of the asthma phenotype or the potential role of formaldehyde in causing new cases of asthma as opposed to exacerbating existing asthma.

For systemic noncancer effects, the draft IRIS assessment identifies associations of formaldehyde exposure with effects on the immune system, the nervous system, the reproductive system, and development. The committee does not find the evidence to be sufficient to support a causal relationship between formaldehyde exposure and those effects, given the weaknesses of the available evidence. First, the committee agrees that there is evidence indicating possible immune effects, including increased incidence of upper respiratory tract infections, respiratory burst activity in the immune system, and modulation of responses to known allergens, but the evidence is insufficient to conclude that these relationships are causal. Second, the committee finds that the draft IRIS assessment overstates the evidence in concluding that formaldehyde is neurotoxic; the selected studies are not sufficient for neurotoxicity-hazard identification, given deficiencies in study design. Third, although the draft IRIS assessment concludes that the epidemiologic studies provide evidence of a convincing relationship between formaldehyde exposure and reproductive and developmental effects, the committee concludes that the evidence indicates a suggestive, rather than convincing, relationship.

The committee supports EPA's selection of the following health effects on which to base a candidate RfC: sensory irritation (eye, nose, and throat), upper respiratory tract pathology, decreased pulmonary function, increased asthma and allergic sensitization, and, despite the weak evidence of causality, reproductive and developmental toxicity. However, as described in Chapters 4 and 5 of the present report, the draft IRIS assessment has substantive problems that weaken the arguments related to those outcomes.

## Selection of Critical Studies

The draft IRIS assessment characterizes the approach for study selection for noncancer outcomes as follows: "in general, studies are included where study quality and ability to define exposures are considered adequate for RfC derivation. Whenever possible, greater consideration is typically given to human data from observational epidemiology studies for derivation of an RfC" (EPA 2010b, p. 5-3). The committee views the stated overall approach as reasonable but found no explicit criteria for its application. Similarly, the concept of "adequate," which appears central in decision-making, is left undefined. The draft IRIS assessment offers six general points that were used to evaluate studies: study size, whether the study evaluated humans or animals, whether an epidemiologic study was in a residential or occupational setting, whether children were included in the study population in a human study, the accuracy of formaldehyde concentration measurements, and whether the study evaluated low formaldehyde concentrations and sensitive end points. The committee agrees that those criteria are appropriate for study selection but notes that no explicit judgments are identified in the draft assessment about how well the individual studies met the criteria. The effects, studies, and points of departure advanced by EPA for candidate RfCs are summarized in Table 6-1. The committee's comments on the studies selected for the specific outcomes are provided in Chapters 4 and 5 of the present report.

**TABLE 6-1** Derivation of Candidate RfCs by EPA[a]

| End Point | Study | Selected POD (range of POD) | $UF_L$ | $UF_S$ | $UF_H$ |
|---|---|---|---|---|---|
| *Respiratory Effects, Asthma, and Sensitization* | | | | | |
| Asthma incidence | Rumchev et al. (2002) | NOAEL, 33 ppb (24-39 ppb) | 1 | 3 | 1 or 3 |
| Increased asthma | Garrett et al. (1999) | LOAEL, 28 ppb (16-41 ppb) | 3 | 1 | 1 or 3 |
| Pulmonary function—reduction in PEFR in children (10%) | Krzyzanowski et al. (1990) | $BMCL_{10}$, 17 ppb ($BMC_{10}$, 27 ppb) | 1 | 1 | 3 |
| *Sensory Irritation* | | | | | |
| Eye irritation, burning eyes | Hanrahan et al. (1984) | $BMCL_{10}$, 70 ppb (LOAEL, >100 ppb) | 1 | 1 | 1 or 3 |
| | Liu et al. (1991) | LOAEL, 95 ppb (70-120 ppb) | 3 | 1 | 1 or 3 |
| | Ritchie and Lehnen (1987) | NOAEL, 50 ppb (0-100 ppb) | 1 | 1 | 1 or 3 |
| *Reproductive and Developmental Toxicity* | | | | | |
| Decreased fecundability density ratio[b] | Taskinen et al. (1999) | NOAEL, 86 ppb (estimated 8-hr TWA) | 1 | 1 | 10 |

[a]All uncertainty factor values are those assigned by EPA. Source: Adapted from Table 5-6 in EPA (2010b).
[b]Decreased fecundability density ratio is estimated as "the conception rate for exposed women relative to that for unexposed women in each menstrual cycle of unprotected intercourse" (Rowland et al. 1992).
Abbreviations: $UF_L$, uncertainty factor for adjustment of LOAEL to NOAEL; $UF_S$, uncertainty factor for adjustment of less than chronic study to chronic duration; $UF_H$, uncertainty factor that accounts for human population variability; RfC, reference concentration; POD, point of departure; EPA, Environmental Protection Agency; PEFR, peak expiratory flow rate; LOAEL, lowest observed adverse effect level; NOAEL, no-observed-adverse-effect-level; $BMCL_{10}$, lower 95% confidence limit on the benchmark concentration corresponding to a 10% response level; and TWA, time-weighted average.

Candidate RfCs were derived for the related group of effects occurring in the respiratory system by using three observational epidemiologic studies of children exposed in their homes. The committee agrees with EPA's assessment that two of the selected studies are sufficient to support derivation of candidate RfCs for decreased pulmonary function (Krzyzanowski et al. 1990) and prevalence and severity of allergic sensitization and respiratory symptoms (Garrett et al. 1999). However, the committee does not support the selection of the Rumchev et al. (2002) study because the end point of "incident asthma" is not supported by an understanding of the phenotype of asthma in the age range of participants in the study.

Candidate RfCs were derived for sensory irritation of the eye by using three residential epidemiologic studies: Hanrahan et al. (1984), Ritchie and Lehnen (1987), and Liu et al. (1991). Although the committee agrees with EPA's selection of the Hanrahan et al. (1984) and Liu et al. (1991) studies as the best of those available, it disagrees with EPA's selection of the Ritchie and Lehnen (1987) study because of the high potential for selection bias among the self-selected participants.

A candidate RfC was derived for reproductive effects on the basis of a decreased fecundability density ratio observed in women occupationally exposed in the epidemiologic study by Taskinen et al. (1999). The committee agrees that the choice of that outcome in the study is justifiable for the reasons provided in the draft IRIS assessment.

The committee supports EPA's decision not to derive candidate RfCs for immunotoxicity and neurotoxicity end points but disagrees with its decision not to calculate a candidate RfC for upper respiratory tract pathology. Many well-documented studies have reported the occurrence of upper respiratory tract pathology in laboratory animals, including nonhuman primates, after inhalation exposure to formaldehyde. The dataset is one of the most extensive available, and the committee therefore recommends that EPA use the animal data to calculate a candidate RfC for this end point.

**Uncertainty Factors**

As defined by EPA (1994, 2010a), uncertainty factors are used to derive an RfC to account for study limitations, uncertainty in required extrapolations, and variability in response:

- $UF_A$ accounts for uncertainty in animal-to-human extrapolation.
- $UF_H$ accounts for human population variability and uncertainty in estimation of the variability.
- $UF_L$ adjusts a lowest observed-adverse-effect level (LOAEL) to a no-observed-adverse-effect level (NOAEL).
- $UF_S$ adjusts a less than chronic study to a chronic duration.
- $UF_D$ accounts for uncertainty in identifying the critical effect when the database does not evaluate a complete array of health effects.

The default value for each uncertainty factor is 10; a factor of 3 (the approximate square root of 10) is used by convention when there is information to support a partial reduction in the uncertainty factor (EPA 1994). Guidance from EPA on when a specific uncertainty factor might be changed from the default value has been provided by the Toxicology Working Group of the 10X Task Force (EPA 1999) and EPA's RfD/RfC Technical Panel (EPA 2002). As noted above, EPA requested advice from the committee on determining the values of the uncertainty factors that account for human population variability and database completeness.

EPA selected study-specific uncertainty factors for each of the candidate RfCs (Table 6-1). All candidate RfCs advanced by EPA are based on observational epidemiologic studies; thus, $UF_A$ that accounts for uncertainty in animal-to-human extrapolation is assigned a value of 1. The committee concurs with EPA's selection of a value of 3 for $UF_L$ for the Garrett et al. (1999) and Liu et al. (1991) studies. Although the studies did not report the duration of residence in the homes tested, the exposure period was assumed to correspond to a chronic exposure period of 10% of a lifetime, or 7 years, as defined by EPA. Thus, EPA selected a value of 1 for $UF_S$ for all studies except Rumchev et al. (2002) for which a value of 3 was selected because the study participants were exposed for less than 3 years.

Two alternative values (1 and 3) are presented in the draft IRIS assessment for $UF_H$ in five of the seven studies for which candidate RfCs were developed (see Table 6-1). In defining $UF_H$, EPA specifically considered susceptible populations, including children, and that is consistent with the NRC (1993) report *Pesticides in the Diets of Infants and Children*, the Food Quality Protection Act (1996), and the EPA (2006) report *A Framework for Assessing Health Risk of Environmental Exposures to Children*. As noted previously, the committee does not support the use of the Rumchev et al. (2002) and Ritchie and Lehnen (1987) studies for derivation of candidate RfCs. Thus, the focus of the remainder of this discussion will be on uncertainty factors used to derive candidate RfCs for asthma and allergic sensitization (Garrett et al. 1999) and eye irritation (Harahan et al. 1984; Liu et al. 1991).

### Evaluation of Human Population Variability

Variability of the human response to a specific exposure is recognized quantitatively during the development of the RfC through application of the uncertainty factor $UF_H$ (EPA 1994). An overarching difficulty in determining the appropriate value for $UF_H$ is that the critical underlying parameters—the proportion of the population to be protected by an RfC and the definition of appreciable risk—have not been quantitatively articulated by EPA or other risk managers. In fact, the definition of an appreciable risk is a societal matter, and the selected value might depend on the particular material of concern and the context (Lowrance 1976; NRC 2009). Furthermore, it is often difficult to determine an appropriate value for $UF_H$ because chemical-specific information on mode of action and on characteristics of the sensitive populations is typically sparse. Consequently, descriptions of human variability are often highly imprecise and uncertain.

$UF_H$ is conceptualized as accounting for population variability that arises from differences in toxicokinetics (variation in the dose to the active site from the same external exposure) and from differences in toxicodynamics (variation in response to the same dose at the active site) (EPA 1994, 2002). Accordingly, the committee evaluated the data presented in the draft IRIS assessment on toxicokinetics, toxicodynamics, mode of action, and attributes of the key studies to consider how well they represent the dose-response data on susceptible populations. The committee found the discussion of potential sources of population variability and uncertainties related to life stages and mode of action in Section 4.6 of the draft assessment to be generally comprehensive. However, sources of uncertainty and variability identified in that section are not integrated into the discussion of the appropriate value of $UF_H$ to use with the key studies; instead, the section focuses primarily on the attributes of the study for the specific candidate RfC. The following sections represent the committee's synthesis of the available information and the response to its charge question.

**Toxicokinetics**

The toxicokinetics of inhaled formaldehyde depend on uptake at the portal of entry and metabolism. Total uptake in the upper respiratory tract might vary from person to person because of different physical characteristics of the upper respiratory tract, breathing patterns (oral vs nasal), and ventilation rate. As noted by EPA, modeling of reactive-gas uptake by Ginsberg et al. (2005) suggests that uptake in the upper respiratory tract is similar in 3-month-old children and adults. That relationship was confirmed by Ginsberg et al. (2010) after reanalysis of the models that used the higher ventilation rates in children reported in the updated *Child-Specific Exposure Factor Handbook* (EPA 2008a). EPA evaluated the computational-fluid-dynamics model of Garcia et al. (2009) that models flux (rate of gas absorbed per unit surface area of the nasal lining) of a generic reactive water-soluble gas, which is representative of formaldehyde, in the individual nasal cavities of five adults and two children, 7 and 8 years old (EPA 2010b, Appendix B). Garcia et al. (2009) report that their simulations of localized flux across the nasal epithelium do not predict differences in nasal dosimetry (uptake) between children and adults; average uptake differed by a factor of 1.6 among the seven subjects. Variability in the local gas flux among different regions of the individual nasal passages of the five adults and the two children was larger (a factor of about 3-5). If the effects associated with formaldehyde exposure are specific to location and cell type in the upper respiratory tract, the variability in local flux could be a contributor to variability in population response. EPA concluded and the committee agrees that the analysis of interindividual flux, although well done, is based on a small sample and involves people whose nasal cavities have a "normal" shape. Consequently, the study probably did not capture the full array of nasal-cavity geometry, and the findings should be generalized with caution. The committee encourages EPA to continue to evaluate the type of data that can aid in characterizing variability in deposited dose in future IRIS assessments.

Ventilation rate is another potential contributor to population variability in toxicokinetics and needs to be evaluated because children have higher ventilation rates in relation to body weight than do adults. Unlike the oral reference dose, the inhalation RfC is typically used directly without adjustment for differences in exposure conditions (EPA 2009a). As noted by EPA, ventilation rate and age-related variation in oral and nasal breathing patterns probably contribute to variability in dose to specific areas of the upper respiratory tract; higher ventilation rates and oral breathing decrease absorption of formaldehyde in the nasal cavity and increase the amount of formaldehyde available to the lower respiratory tract (EPA 2010b).

As described in the draft IRIS assessment, formaldehyde is metabolized primarily by alcohol dehydrogenase (ADH3) (EPA 2010b). ADH3 plays a central role in regulating bronchiole tone and allergen-induced hyperresponsiveness (Gerard 2005; Que et al. 2005) and mediates reduction of $S$-nitrosoglutathione (GSNO) (Thompson and Grafstrom 2008; Thompson et al. 2010), an endogenous bronchodilator and reservoir of nitric oxide activity (Jensen et al. 1998). The ontogeny and regulation of ADH3 among human life stages is not yet understood (Thompson et al. 2009). ADH3 mRNA transcripts have been detected in the third-trimester human fetus, but the relative expression and activity of ADH3 protein at various life stages are not known (Thompson et al. 2009). Polymorphisms in ADH3 have been reported in members of various ethnic groups (Hedberg et al. 2001), and single-nucleotide polymorphisms in ADH3 have been associated with childhood risk of asthma (Wu et al. 2007). As noted in the draft IRIS assessment, the qualitative and quantitative effects of the interactions of ADH3 and GSNO on the toxicity of formaldehyde and human population variability are not understood.

**Toxicodynamics**

Toxicodynamics is a potential source of human population variability related to variation in the response to a given dose at the active site. The potential contribution of toxicodynamic differences to population variability is evaluated by considering the mode of action, potential life-stage sensitivities, and the extent to which the study population includes susceptible populations. Although the modes of action of formaldehyde's effects on the respiratory system are not fully characterized, the committee finds the discussions of the biologic mechanisms underlying sensory irritation, asthma, and immunotoxicity in the draft IRIS assessment to be inadequate and not reflective of current scientific understanding. Formaldehyde has been shown to activate the TRPA1 ion channel irreversibly by covalent modification (Macpherson et al. 2007). The TRPA1 ion channel is associated with sensory irritation responses (Bessac and Jordt 2008) and plays a critical role in allergic asthmatic responses as a major neuronal mediator of allergic airway inflammation (Caceres et al. 2009). The contribution of TRPA1 and the enzymes involved in metabolism or processing of formaldehyde—ADH3 (Gerard 2005; Que et al. 2005; Wu et al. 2007; Hedberg et al. 2001) and GSNO (Thompson and Grafstrom 2008)—to population variability in toxicodynamics is not understood.

Populations sensitive to effects of formaldehyde exposure include those who have asthma (Krzyzanowski et al. 1990; Kriebel et al. 1993; Garrett et al. 1999) and atopy (Garrett et al. 1999). They may also include those who have acute and chronic inflammatory airway conditions (such as viral infections, asthma, rhinitis, and chronic obstructive pulmonary disease) (Bessac and Jordt 2008) and those exposed to other respiratory irritants that act through related modes of action (Macpherson et al. 2007; Bessac and Jordt 2008). Children may be a susceptible population, given their developing respiratory tract and nervous system (Pinkerton and Joad 2000; Rice and Barone 2000; Ginsberg et al. 2005).

On the basis of the toxicokinetic and toxicodynamic data, the committee agrees with EPA's conclusion that the available data are consistent with some life-stage differences in susceptibility to the effects of formaldehyde. However, there is substantial uncertainty regarding the determinants and the distribution of susceptibility in the population.

**Values of $UF_H$**

The committee considered the appropriate value for $UF_H$ for the following studies: Garrett et al. (1999), which evaluated the risk of allergy and asthma-like respiratory symptoms in 148 children 7-14 years old; Liu et al. (1991), which evaluated eye irritation in over 1,000 people 4 to over 65 years old; and Hanrahan et al. (1984), which evaluated eye irritation in 61 teens and adults. Criteria described by the RfD/RfC technical report (EPA 2002) regarding when a value of less than 10 could be assigned to $UF_H$ guided the committee in its assessment of the appropriate value for $UF_H$ (1 or 3). Specifically, "how completely the susceptible subpopulation has been identified and their sensitivity described (vs. assumed)" and whether "the data set on which the POD [point of departure] is based is representative of the exposure/dose-response data for the susceptible subpopulation(s)" (EPA 2002, p. 4-43, 4-44).

*Identification of Sensitive Populations*

Children and adults who have asthma and allergic sensitization are susceptible populations on the basis of studies that showed increased exacerbation of respiratory and allergic sensitization responses to formaldehyde exposure in people who have asthma (EPA 2010b, p. 4-543). Increased symptoms of upper airway irritation were observed in study participants that also reported chronic respiratory and allergy symptoms; this finding suggests increased susceptibility to irritation (Liu et al. 1991). Subjects who have acute and chronic inflammatory airway conditions (such as viral infections, asthma, rhinitis, and chronic obstructive pulmonary disease) (Bessac and Jordt 2008) may also be susceptible populations. However, the mode of action for formaldehyde's effects is not sufficiently elucidated to understand the influence of such factors as life stage, respiratory tract development, latency, underlying disease status (such as chronic respiratory diseases and allergic symptoms), genetic polymorphisms of ADH3 and aldehyde dehydrogenase, and cumulative effects of exposure to chemicals that affect the same targets as formaldehyde.

To support a value of 1 for $UF_H$, EPA cites the RfD/RfC technical report, which indicates that a $UF_H$ of 1 has been applied in cases in which data are very specific "about the particular vulnerability of infants and children within specific age ranges to an agent" (EPA 2002, p. 4-43). To determine how often EPA has used a $UF_H$ of 1 in derivation of reference values and its underlying rationale, the committee searched the IRIS database and identified six RfDs with a value of 1 assigned for $UF_H$ (EPA 2010c). The RfDs are those for benzoic acid, beryllium, fluorine, manganese, nitrate, and nitrite.[1] In contrast with formaldehyde, for example, the RfDs for nitrate and nitrite identified points of departure from studies of the susceptible population (infants) and noted that the duration of susceptibility to the effects of nitrate is short (that is, children are not susceptible after specific points in development are reached). In the view of the committee, the modes of action for formaldehyde effects on the respiratory tract are not sufficiently understood to determine all potential susceptible populations, and the factors contributing to susceptibility are not yet adequately described. Thus, the committee does not support the application of a value of 1 for $UF_H$.

*Representativeness of Exposure and Dose-Response Data*

For the candidate RfC for asthma and allergic sensitization that was based on the study by Garrett et al. (1999), the draft IRIS assessment assumes that children and adults who have asthma or allergic sensitization are the susceptible populations. As described by EPA, the Garrett et al. (1999) study includes a higher proportion of children that may be predisposed to asthma and allergic sensitization than is found in the general population (53 of the 148 children in the study had a diagnosis of asthma); thus, the study appears to describe responses in susceptible populations (EPA 2010b). Garrett et al. (1999) reported that the children who were most responsive to the effects of formaldehyde had parents or family members who had asthma or atopy; this lends support to the hypothesis that there is a genetic component to the increased sensitivity of these children, but there could also be unrecognized environmental sources that contribute to similarities in responsiveness within families.

---

[1]The IRIS database was searched to identify RfCs and RfDs that were derived by using a value of 1 for $UF_H$. A search for a $UF_H$ of 1 yielded no results. A search for a composite UF of 1 yielded five chemicals for which a $UF_H$ of 1 was used to derive RfDs: benzoic acid, fluorine, manganese, nitrate, and nitrite. A search for a composite UF of 3 yielded no results. A search for a composite UF of 10 yielded one chemical (beryllium) for which a $UF_H$ of 1 was used to derive an RfD.

For the candidate RfC for sensory irritation, the draft IRIS assessment does not identify a potentially susceptible population but notes that the studies by Liu et al. (1991) and Hanrahan et al. (1984) were population-based. The Liu et al. (1991) study was large; it included children less than 4 years old, the elderly, and both sexes and reported the highest prevalence of eye irritation in participants 20-64 years old (EPA 2010b, p.5-60). Of the study population, 33% reported pre-existing respiratory conditions, including allergy, asthma, chronic bronchitis, and emphysema (Liu et al. 1991). The small number of people (61) and the absence of young children lessen the confidence that the Hanrahan et al. (1984) study is sufficiently representative of a sensitive population. However, as noted in the draft IRIS assessment, the exposure-response (prevalence) relationship is similar in the two studies (EPA 2010b, p.5-60).

The inclusion of potentially susceptible populations in the studies supports a reduction of $UF_H$ from the default value of 10 for the candidate RfCs on the basis of the Garrett et al. (1999), Liu et al. (1991), and Hanrahan et al. (1984) studies. EPA has long used the square root of the default value of 10 when reducing uncertainty factors from the default value; thus, the committee supports a value of 3 for $UF_H$ for the candidate RfCs on the basis of the Garrett et al. (1999), Liu et al. (1991), and Hanrahan et al. (1984) studies.

## Evaluation of Database Completeness

As noted above, the committee was asked to comment on appropriate values for the uncertainty factor that accounts for database completeness, $UF_D$. The draft IRIS assessment presents several options for $UF_D$: (1) apply a $UF_D$ of 1 with a qualification that further research on reproductive, developmental, and neurotoxic effects would be valuable; (2) apply a $UF_D$ of 1 with a qualification that the RfC is explicitly protective; (3) apply a $UF_D$ of 3; or (4) provide two RfCs, one with a $UF_D$ of 1 that is protective of the better-studied effects and a second with a $UF_D$ of 3 to account for the limitations of the data on reproductive, developmental, and neurotoxic effects (EPA 2010b, p. 5-72).

An RfC is derived on the basis of the critical effect (the effect that occurs at the lowest exposure) and is intended to provide protection from all noncancer effects. The final question in the process is "Do the results of all the studies indicate the possibility of effects on particular systems that have not yet been explored sufficiently or do they indicate that additional studies may reveal effects not yet characterized?" (EPA 2002, p. 4-21). One must consider effects across all life stages from conception to old age, subtle effects that affect a person's quality of life, and effects that may occur after a long latency period (EPA 2002). The database on formaldehyde is extensive and includes evaluation of health effects in the human population. EPA evaluated a broad array of health effects associated with formaldehyde exposure, including those related to asthma, pulmonary function, sensory irritation, respiratory tract pathology, reproductive toxicity, developmental toxicity, neurotoxicity, and immunotoxicity.

For derivation of the RfC, the draft IRIS assessment selects critical effects related to the respiratory system: 10% reduction in peak expiratory flow rate in children at a $BMCL_{10}$ of 17 ppb (Krzyzanowski et al. 1990), increased prevalence and severity of allergic sensitization and increased severity of respiratory symptoms in children with a LOAEL of 28 ppb (Garrett et al. 1999), and "incidence of asthma" (Rumchev et al. 2002). As indicated above, the committee does not recommend the use of Rumchev et al. (2002) to derive the RfC. The proposed RfC is thus based on subtle effects observed in children who are expected to be a susceptible population for respiratory effects.

As discussed in the draft IRIS assessment, the principal deficiencies in the database are the lack of studies that provide a full evaluation of the complete spectrum of end points in the reproductive, developmental, nervous, and immune systems and the absence of a multi-generation animal study that evaluates reproductive function. The committee notes a critical need for epidemiologic studies that have high-quality exposure data to examine associations with potential effects. Some information, however, is available for evaluating each system noted by EPA. Reproductive effects were evaluated in women occupationally exposed to formaldehyde; the Taskinen et al. (1999) study provided a NOAEL of 86 ppb (adjusted concentration) for decreased fecundity density ratio. Animal developmental studies and male reproduction studies evaluated effects at concentrations of 10 ppm and higher with little evidence of adverse developmental or reproductive effects in the absence of overt signs of toxicity. EPA considered the rat study of Malek et al. (2003a) with a LOAEL of 100 ppb (adjusted concentration) based on performance of learning tasks to provide evidence of neurotoxicity in animal studies. However, the committee found the quality of the study designs used in this and other available studies to be deficient for neurotoxicity hazard identification. Effects related to the immune system were observed in an occupational cohort and in children. Increased susceptibility to upper respiratory tract infections was observed in the occupational study by Lyapina et al. (2004) with a LOAEL of 722 ppb. Increased prevalence and severity of allergic sensitization in children was observed in the study by Garrett et al. (1999) with a LOAEL of 28 ppb, which was used as the point of departure for the RfC proposed by EPA.

The RfC is based on respiratory effects evaluated in children, including children who may be more susceptible than the general population of children. Although there are gaps in the data, as noted above, the database provides information on effects in various systems (reproductive, developmental, immune, and nervous systems) that are of special concern for identifying effects in sensitive populations. Health effects in those systems were observed at exposures higher than those at which effects were observed in the respiratory system. However, NOAELs were not identified for most effects; as noted by EPA, the lack of clear NOAELs contributes to the uncertainty as to whether the RfC would be protective for those health effects (EPA 2010b, p. 6-29).

Determining the value for $UF_D$ is difficult because it is always challenging to predict what is not known. The formaldehyde database is extensive but has some gaps. However, the breadth of the database suggests that it is unlikely that effects will be observed in organ systems not already identified as affected by formaldehyde. The difficult question to answer is, Do the results of the studies "indicate that additional studies may reveal effects not yet characterized?" (EPA 2002, p. 4-21). A quantitative framing of the question is useful to focus the decision-making process, that is, What is the likelihood that effects not yet studied could occur at exposures lower than the known effects, and if new effects occur, how much lower could those exposures be? (Evans and Baird 1998). As described above, the effects of formaldehyde on the reproductive, developmental, nervous, and immune systems are not completely characterized, and the modes of action are not known. It is possible that better studies could reveal effects that occur at exposures lower than those currently evaluated in those systems. However, the draft RfC is based on effects at the portal of entry on the respiratory system that are observed in children who are thought to be particularly sensitive to the effects of formaldehyde. Thus, the likelihood that as yet unstudied effects occur at exposures lower than those currently used as the basis of the draft RfC appears low. Accordingly, the committee recommends that EPA adopt its first option and apply a $UF_D$ of 1 with a qualification that further research on reproductive, developmental, neurotoxic, and immunotoxic effects would be valuable.

**Comments on the IRIS Process for Deriving Reference Concentrations**

The draft IRIS assessment develops several candidate RfCs in accordance with the recommendations of EPA's guidelines (2002). However, there is little synthesis of the relationships among the health effects identified in the respiratory tract and among target organs in the draft IRIS assessment until the final summary after RfC derivation in Section 6.1.3. Each respiratory tract end point (sensory irritation, upper respiratory tract pathology, decreased pulmonary function, increased asthma, and allergic sensitization) and its associated dose-response information are considered individually, although the draft assessment acknowledges that they are etiologically and clinically related. Thus, the draft assessment appears excessively driven by the need to identify the best study to represent each end point at the cost of overlooking studies that identify related effects at slightly higher concentrations or animal studies that would inform the biologic exposure-response story.

A clearer presentation of information with more tables that summarize available studies, figures that synthesize related effects from multiple studies (see Figure 6-2), and greater integration of information about mode of action and potentially susceptible populations during study selection and assignment of uncertainty factors would improve the assessment's ability to make a compelling case for the RfC ultimately put forward.

The approach described and illustrated in Figure 6-2 has been used in recent EPA assessments of tetrachloroethylene (EPA 2008b) and trichloroethylene

(EPA 2009b). The Science Advisory Board, which conducted a peer review of EPA's draft IRIS assessment titled *"Toxicological Review of Trichloroethylene,"* commented that "the Panel supported the selection of an RfC and an RfD based on multiple candidate reference values in a narrow range, rather than basing these values on the single most sensitive critical endpoint. This approach was supported by the Panel because it was a very robust approach that increases confidence in the final RfC and RfD" (EPASAB 2011, p. 39). The National Research Council (NRC) Committee to Review EPA's Toxicological Assessment of Tetrachloroethylene also offered advice on the graphic presentation of reference values as part of the noncancer assessment. It was noted that "the committee strongly supports the use of … graphical aids… to make it clear which uncertainty factors were applied, to which studies they were applied, and the effects of particular assumptions" (NRC 2010, p. 93).

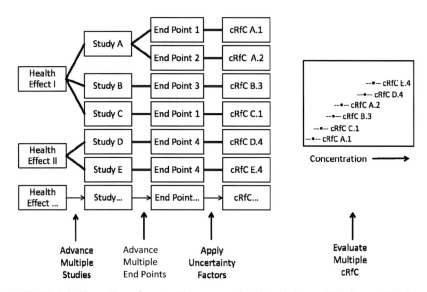

**FIGURE 6-2** Illustration of a potential process for identifying an RfC from a full database. Health effects in organ systems associated with exposure to the chemical are identified. For each health effect, studies that meet criteria for inclusion are advanced. From each study, one or more end points that meet specified criteria are advanced, and the point of departure is identified and adjusted to a human-equivalent concentration. Uncertainty factors are selected on the basis of study and end-point attributes and applied to the point of departure to yield candidate RfCs (cRfCs). All cRfCs are evaluated together with the aid of graphic displays that incorporate selected information on attributes relevant to the database and the decision to be made. A final RfC is selected from the distribution after consideration of all critical end points from studies that met the criteria for inclusion.

The committee concurs that appropriate graphic aids that enable the visualization of the range of concentrations evaluated in each published study selected for quantitative assessment may help to identify clusters of studies and especially low or high reference values that may not agree well with the body of literature. The NRC Committee to Review EPA's Toxicological Assessment of Tetrachloroethylene, argued that "the convergence of sample reference values into clusters would confer confidence on the use of a critical study if other studies led to similar conclusions" and that "convergence of estimated values from studies that are methodologically sound, even if they are not listed as key, would support the RfC proposed by EPA" (NRC 2010, p. 93). That committee's arguments are highly relevant to the risk assessment of formaldehyde.

## FORMALDEHYDE UNIT RISKS FOR CANCER

Sections 5.2 and 5.3 of the draft IRIS assessment provide detailed accounts of the derivation of unit risks for formaldehyde inhalation and also discuss sources of uncertainty and variation in the estimates. Briefly, EPA chose as the primary study the National Cancer Institute (NCI) occupational cohort of U.S. workers involved in the production or use of formaldehyde (Blair et al. 1986). Within the cohort, EPA identified nasopharyngeal cancer (NPC) as the primary cancer associated with formaldehyde in a follow-up of the NCI study to 1994 (Hauptmann et al. 2004) and also selected leukemia and Hodgkin lymphoma as additional cancers to evaluate on the basis of an extended follow-up of the same cohort to 2004 (Beane-Freeman et al. 2009). Relying on the Poisson regression models for cancer mortality reported in Hauptmann et al. (2004) and Beane-Freeman et al. (2009), which used a detailed workplace exposure inventory, EPA estimated life time risks (probabilities) of cancer mortality by using a life-table analysis in conjunction with mortality and cancer incidence in the U.S. population and estimated effective concentration (EC) exposure to formaldehyde corresponding to 0.5-0.05% extra risk. It then used linear extrapolation from the upper confidence limit of the EC to derive cancer unit risk estimates for the three cancers separately. It concluded that the unit risk is in the range of $1.1 \times 10^{-2}$ to $5.7 \times 10^{-2}$ $ppm^{-1}$ for a single cancer group and has an upper bound of $8.1 \times 10^{-2}$ $ppm^{-1}$, which combines the risks of the three cancers. Analyses conducted by EPA that used animal data on the incidence of nasal squamous cell carcinoma in F344 rats yielded a human-equivalent unit risk of $1.2 \times 10^{-2}$ to $2.2 \times 10^{-2}$ $ppm^{-1}$. The draft IRIS assessment concludes that $8.1 \times 10^{-2}$ $ppm^{-1}$ is a reasonable estimate of unit risk of total cancer.

Many sources at various stages of the risk-estimation process contribute to the overarching uncertainty and variation in the final risk estimates. The effect of the uncertainties in aggregation is complex and difficult to quantify. The draft IRIS assessment makes commendable efforts in discussing a number of sources of uncertainties. The committee's appraisal of EPA's analyses follows and focuses on some key factors that determine the overarching uncertainty in cancer unit risk estimates, including the choice of study, cancer end point, dose metric, dose-response model, point of departure, and extrapolation to low doses.

## Selection of Studies

EPA reviewed a number of cohort studies for the purpose of estimating a unit risk for formaldehyde inhalation and chose the studies of the NCI cohort (Hauptmann et al. 2004; Beane-Freeman et al. 2009). EPA's choice has the following basis: the NCI studies used the largest known cohort, had detailed individual exposure estimates to support dose-response assessment, and used an internal comparison group that is less likely to be confounded by the "healthy-worker" effect than is an external reference group. Results from two follow-ups of the NCI cohort were selected. The first (Hauptmann et al. 2004) went up to 1994 and included a total of 865,708 person-years in 10 U.S. plants, and the second (Beane-Freeman et al. 2009) went through 2004 and had a total of 998,106 person-years of follow-up. More important, the NCI studies exhibited positive and statistically significant exposure-response relationships for some exposure metrics and selected cancers, notably NPC, Hodgkin lymphoma, and leukemia. EPA also reviewed other epidemiologic studies but chose not to use them for unit risk estimation. EPA judged the results from the other studies to be consistent with those of the NCI studies but concluded that various limitations prevented them from being used for quantitative risk estimation, such as a lack of sufficient quantitative exposure data to permit dose-response assessment. Chapters 4 and 5 of the present report provide further discussion of EPA's study selection.

The committee agrees that the NCI studies are a reasonable choice because they are the only ones with sufficient exposure and dose-response data for risk estimation. However, the NCI studies have limitations. The committee is concerned about the clustering of seven of nine NPC deaths in a single plant (Hauptman et al. 2004) and missing death reports (Beane-Freeman et al. 2009). The committee strongly encourages EPA to state its inclusion and exclusion criteria clearly for its systematic review, analysis, and selection of studies. Systematic use of such criteria enhances the transparency of risk assessment.

In principle, identifying the "best" study for general risk-assessment purposes is neither feasible nor necessary. Inclusion of multiple studies that meet the selection criteria will enhance EPA's ability to examine variability and uncertainty attributable to, for example, different study designs, populations, and exposure conditions.

## Selection of Cancer End Points

NPC was the only respiratory cancer in Hauptmann et al. (2004) that exhibited a positive exposure-response relationship for cancer mortality under all four dose metrics (peak, average intensity, duration, and cumulative exposure).

The exposure-response relationship was statistically significant under the metric of peak exposure, was marginally significant under cumulative exposure and average intensity, and not significant for duration. The draft IRIS assessment notes that prostate cancer showed a statistically significant trend with peak exposure and that bone cancer was associated with peak and cumulative exposure. However, EPA decided to exclude those two cancers from further assessment, and the committee agrees with that decision.

On the basis of the findings from the extended 2004 follow-up of the NCI cohort, EPA identified Hodgkin lymphoma and leukemia as two additional cancers for which to estimate unit risk (Beane-Freeman et al. 2009). Hodgkin lymphoma (27 deaths) showed a positive, statistically significant exposure-response relationship in cancer mortality for peak exposure but was only marginally significant for average intensity and cumulative exposure. The exposure-response trend of leukemia mortality (123 deaths) was only marginally significant for peak and cumulative exposure and was not significant for average intensity (Beane-Freeman et al. 2009).

The lack of consistency in exposure-response relationships between various exposure metrics and the three types of cancer is of concern. The inconsistency may simply be a result of applying multiple metrics, some of which are not highly valid or precise or are perhaps less relevant to the underlying mechanisms. It could also reflect the absence of causal mechanisms associating, for example, leukemia with formaldehyde exposure.

The committee agrees that EPA's choice of NPC, Hodgkin lymphoma, and leukemia to estimate the unit risk is appropriate given that the use of Hodgkin lymphoma and leukemia primarily supports the assessment of uncertainty and the magnitude of cancer risk where there is a lack of evidence to support the biologic plausibility of a relationship between formaldehyde exposure and the two cancers. The committee also notes that the positive exposure-response findings from the NCI studies support the use of the three cancers for unit risk estimation. However, there are major uncertainties in using the cancers for risk estimation. As discussed in Chapter 5 of the present report, there is a noticeable lack of evidence of a causal relationship of formaldehyde exposure and Hodgkin lymphoma or leukemia. In contrast, there is strong epidemiologic evidence of a causal relationship of formaldehyde exposure and NPC. However, that seven of nine NPC deaths occurred in the Wallingford, Connecticut, factory in the NCI cohort is intriguing. Marsh et al. (2002) investigated the Wallingford cohort, 7,328 workers who were employed in the factory during 1941-1984. Follow-up continued until 1998, and exposure history was reconstructed independently of Marsh et al. (2002). Although the analysis of Marsh et al. (2002) appears to support an increased standardized mortality ratio (SMR) in association with categories of increased formaldehyde exposure, the fact that the SMR was similar in workers with exposure history of less than 1 year (SMR, 5.35; 95% CI, 1.46-14; 4 deaths) and exposure history of greater than 1 year (SMR, 4.59; 95% CI, 0.95-13; 3 deaths) led the authors to question the association with formaldehyde. Marsh and Youk (2005) reanalyzed the NPC data reported in Hauptmann et al.

(2004) and argued that the exposure-response relationship in the NCI cohort was driven largely by the Wallingford, Connecticut, plant. Marsh and colleagues (Marsh et al. 2007a,b) further speculated, through analysis of a case-control study nested within the NCI cohort, that the exposure-response relationship in NPC mortality could be due to the workers whose prior occupation was silversmithing. The committee notes that although data are insufficient to substantiate the confounding argument of Marsh et al. (2007a,b), uncertainties about the causal relationship between formaldehyde exposure and NPC mortality also exist and cannot be eliminated on the basis of the NCI-cohort studies alone even when the data are pooled over all 10 plants. The negative findings in the nine other plants (that is, the plants other than the one in Wallingford, Connecticut) need to be considered.

Although EPA's choice of NPC, Hodgkin lymphoma, and leukemia for its cancer assessment seems defensible in light of dose-response data requirements described in its carcinogenicity risk-assessment guidelines (EPA 2005), it needs to provide a clear description of its criteria for selecting cancers for its assessment and clearly demonstrate a systematic application of the criteria. For example, how strong a dose-response relationship is required for inclusion? How strong does the evidence of a causal relationship need to be? For leukemia mortality, the Poisson regression-based trend test yielded p values of 0.08 and 0.12 for all and exposed person-years only, respectively, in the case of cumulative exposure (EPA 2010b, Table 5-12). However, the draft IRIS assessment states that "only all leukemias combined and Hodgkin lymphoma were judged to have exposure-response data adequate for the derivation for unit risk estimates" (EPA 2010b, p. 5-75). In that case, did the preferred metric of cumulative exposure outweigh the less significant dose-response relationship? Whereas the committee supports the use of multiple cancers to demonstrate sensitivity, variability, and uncertainty in cancer risk estimation, it recommends that EPA describe and systematically apply a set of selection criteria for cancer end points.

**Selection of Dose Metrics**

The NCI cohort studies explored four exposure metrics: peak, average intensity, duration, and cumulative exposure. EPA reasoned that cumulative exposure is preferable for dose-response assessment. In the NCI cohort, peak exposure is based on frequency of periods, typically less than 15 min, in which concentrations were above an 8-hr time-weighted average (TWA). The frequency of such periods of increased concentrations led to the ranking of peak exposure as low, medium, and high. Thus, peak exposure does not fully account for the actual concentrations during the peaks, nor does it account for exposure duration. Average intensity derived from an 8-hr TWA also fails to account for exposure duration. Exposure duration does not account for concentrations that vary over time and work place. Therefore, cumulative exposure that combines duration and average intensity into one measure is a more reasonable and acceptable choice for quantitative exposure-response assessment. However, contrary to the draft IRIS assessment (EPA 2010b, p. 5-79), a statistically significant trend test with cumulative exposure does not necessarily indicate a good fit for the cumulative exposure. The committee agrees with EPA about the choice of cumulative exposure as the metric for risk estimation but notes that other dose metrics can be biologically applicable in various situations. Peak exposure, for example, can be relevant for effects more likely to be induced by acute and high exposure to formaldehyde.

## Choice of Dose-Response Models

EPA relied on Poisson regression to model exposure-response associations in mortality related to the selected cancers. Poisson regression is commonly used to analyze mortality in follow-up studies (Breslow and Day 1987; Hauptmann et al. 2004). Its strengths include using person-years at risk as a sample unit of risk, using individual workers' data (death and risk profile), the ability to stratify by population characteristics (such as sex, ethnicity, and age), and the ability to incorporate time from initial exposure to death or cancer occurrence. The Poisson regression models were initially reported by Hauptmann et al. (2004) for NPC mortality and by Beane-Freeman et al. (2009) for leukemia and Hodgkin lymphoma mortality. The logarithm of mortality rate or SMR is modeled as a linear function of exposure and other risk factors, and the incidence of death is assumed to follow the Poisson distribution. The draft IRIS assessment calls it a "log-linear model," which in statistics literature commonly refers to methods designed for analyzing cross-tabulated frequency data. EPA used the rate ratio (RR) that results from the Poisson regression models to evaluate the exposure-response relationship in cancer mortality. To conduct risk assessment with the models directly, EPA obtained the regression coefficients of the models from the authors through personal communications because the coefficients were not reported in the published papers. The draft IRIS assessment should provide more details on the models, including, for example, the degree to which the model fits the data.

Conducting Poisson regression for mortality or incidence in follow-up studies may require extensive data manipulation to create person-years at risk according to patterns of all covariates that are being considered. It is difficult for the committee to confirm to what extent EPA tried to validate and verify the published models before using them for risk assessment. Whereas the original models may inform a significant association between formaldehyde exposure and NPC mortality, for example, they may result in a point of departure that is unreliable for low-dose extrapolation if they do not fit the data adequately well at low exposures and thus become inappropriate for risk-assessment purposes. That issue is a concern especially for events that have extremely low rates at which Poisson assumptions may not hold. One may also wonder whether there were any covariates (such as sex) that interacted with formaldehyde exposure.

The presence of any interactions that indicate effect modification will make the extra-risk formula $(R_x - R_0)/(1 - R_0)$ (EPA 2010b, p. 5-81) depend on the covariates involved rather than independent, as assumed in the draft IRIS assessment. Also of concern is whether the log-linear model reflects the true underlying shape of the exposure-response relationship. Therefore, the committee recommends that EPA conduct its own analysis to confirm the degree to which the Poisson models fit the data appropriately, including an evaluation of goodness of fit, potential interactions between covariates and exposure, and nonlinearity. That analysis is essential because dose-response models used for risk estimation must fit the data well in the low-dose range; this may not necessarily be required for an analysis of the dose-response relationship to establish an association between formaldehyde exposure and a selected cancer. EPA is encouraged to consider the use of alternative extrapolation models, including Cox regression models and nonlinear model forms. The details of such modeling activities should be included in an appendix to the IRIS assessment in sufficient detail that the results can be reproduced.

Callas et al. (1998) conducted extensive analyses and simulations, using the 1994 follow-up data of the same NCI cohort. They reported that Poisson regression models can substantially underestimate RR, hence relative risk of deaths from rare cancers, especially at low exposures. For example, they showed that the relative risk estimated with the Poisson model was 22% less than with Cox regression models at exposures of 0.05- 0.5 ppm-year. Consequently, using the Poisson model may lead to considerable underestimation of unit risk. The authors suggested that Cox regression be used when confounding cannot be well controlled or when age at cancer death does not follow an exponential distribution.

Because seven of the nine NPC deaths were from the Wallingford plant alone, a subgroup analysis of the Wallingford plant with Poisson regression would afford a valuable opportunity to understand better the uncertainty associated with the clustering of the deaths. That would require EPA to conduct Poisson regression on the raw data.

The committee recognizes the substantial resources required to conduct alternative and independent analyses of such data as the follow-up of the NCI cohort. However, having alternative models and analyses enhances EPA's ability to quantify variability and uncertainty attributable to models. EPA's carcinogenicity risk-assessment guidelines recommend considering alternative models, especially when biologically based dose-response models are unavailable (EPA 2005). That exercise is especially important when a single study with uncertainties associated with selected cancers and inconsistency with exposure metrics is used.

## Lifetime Risk of Cancer Mortality or Incidence

EPA's process of estimating unit risk appears to consist of the following steps:

(a) Obtain the Poisson regression model parameter estimates that assume no effect modification due to interactions between exposure and other factors and a log-linear relationship.

(b) Convert the RR estimates from the Poisson model into a lifetime (up to 85 years) mortality risk (probability), R(EC), for NPC, leukemia, or Hodgkin lymphoma for any given formaldehyde exposure, EC. That conversion requires the use of the life-table method (EPA 2010b, Appendix C) in conjunction with the Poisson model mortality risk, age-specific all-cause mortality rates in the U.S. population, and NPC mortality rates derived from the NCI Surveillance Epidemiology and End Results (SEER) database. ECs are converted from occupational exposures to continuous environmental exposures to account for differences in the number of days exposed per year (240 vs 365) and the amount of air inhaled per day (10 vs 20 m$^3$).

(c) Compute lifetime risk of NPC death, $R_0$, in an unexposed reference population, using the U.S. age-specific all-cause mortality data and NPC incidence data from SEER.

(d) Use the equation

Extra risk = $[R(EC_x) - R_0]/(1 - R_0)$

to determine $EC_x$ and its lower confidence limit, $LEC_x$. The extra risk was 0.05% for NPC and Hodgkin lymphoma and 0.5% for leukemia. Extra risk is typically 1-10%.

(e) Use $EC_x$ corresponding to x% increase in risk as a point of departure. The unit risk is equal to $x/LEC_x$.

Steps (b) and (c) are complicated, but the draft IRIS assessment provides only the following brief explanation:

> Extra risk estimates were calculated using the β regression coefficients and a life-table program that accounts for competing causes of death. U.S. age-specific 1999 all-cause mortality rates for all race and gender groups combined (National Center for Health Statistics [NCHS], 2002) were used to specify the all-cause background mortality rates in the life-table program. NCHS 1996-2000 age-specific background mortality rates for NPC were provided by Dr. Eisner of NCI's Surveillance Epidemiology and End Results (SEER) program (EPA 2010b, p. 5-81).

Despite an illustration given in Appendix C of the draft IRIS assessment for the computation process of R(EC), the committee finds that additional detail and explanation would be helpful in making the underlying assumptions clear, the process transparent, and uncertainties better understood. For example, EPA should clarify that the hazard rate for NPC mortality in a given age group is the product of the population NPC mortality that is derived from the SEER database and the probability (as approximated by RR) derived from the Poisson model of observing any NPC death in the occupational cohort. Moreover, EPA appears to have used age-specific NPC incidence from SEER to replace age-specific NPC mortality (column D of Appendix C, EPA 2010b). The implication could be an upward inflation of cancer risk because NPC survival rate is high (Lee and Ko 2005). EPA's computation for $R_0$ and R included all groups under 30 years old. The NCI cohort workers were at least 16 years old when occupational exposure to formaldehyde began, and there was a 15-year lag for NPC mortality in the Poisson regression model. Therefore, the risk of NPC death in the NCI cohort before the age of 30 years is essentially ignored in EPA's model; nonetheless, EPA includes groups less than 30 years old in computing $R_0$ and R. Some explanation of the conversion of cumulative exposure (ppm-year) used in the dose-response model and extra risk to average intensity for EC in R also would be helpful.

The committee recognizes the complexity of dose-response and risk-quantification processes for the occupational cohort mortality data and therefore recommends inclusion of adequate description and interpretation to ensure transparency and readability. Specifically, sufficient detail about data, models, methods, and software should be provided in an appendix to any IRIS assessment to allow independent replication and verification.

### Selection of Point of Departure

EPA's carcinogenicity risk-assessment guidelines (EPA 2005) recommend the use of an extra risk of 1-10% for deriving effective concentration, $EC_x$. The recommended range of risk increase is expected to be within the available data range. The draft IRIS assessment makes an unusual choice of 0.05% for NPC and Hodgkin lymphoma. EPA justified the choice on the grounds that NPC death is rare in the general population (background lifetime risk, $2.2 \times 10^{-4}$), so a 1% increase would be well above the observed range of the NCI data and would result in upward extrapolation. The extra risk of 0.05% corresponds to an RR within the model-based RR range for the cohort. If a higher extra risk were used, the uncertainty of low-dose extrapolation would be greater. Given the extreme rarity of NPC and Hodgkin-lymphoma death, EPA's choice of point of departure is reasonable.

### Derivation of Unit Risks by Using Linear Extrapolation

To derive unit risk estimates for formaldehyde inhalation, EPA relied on the default option of low-dose linear extrapolation. EPA justified its choice on several grounds. First, there is a plausible mutagenic mode of action for NPC and other upper respiratory track cancers. Second, the extra risk appeared linear with exposure below 0.01 ppm on the basis of a comparison of risks that were taken directly from the fitted dose-response models at various exposures. However, the committee notes that the computation was driven entirely by the fitted Poisson model; the degree to which the model fits the data on NPC, Hodgkin lymphoma, or leukemia is not verified or documented. Third, there is no well-established mechanistic dose-response model.

Linear extrapolation entails three steps. First, a dose-response model, often a mathematical function in the absence of reliable information on mode of action, that fits the observed data appropriately well within the available data range must be identified. In the present case, it is the Poisson models fitted to NPC, leukemia, and Hodgkin lymphoma mortality rates. It is less clear how the model fits the datasets. Second, a point of departure is determined from the fitted dose-response model that corresponds to an exposure concentration ($EC_x$) that induces a specified risk increase (x) above that of a reference population. EPA chose a point of departure that corresponds to 0.05% extra risk in lifetime NPC mortality with the risk $R(EC_x)$ derived from the Poisson models and the life-table method. Third, the extra risk level is divided by the point of departure ($EC_x$ or $LEC_x$) to yield a unit risk or slope factor.

Applying linear extrapolation to NPC mortality data on exposed workers yields only a unit risk of $5.5 \times 10^{-3}$ based on 0.05% extra risk and $LEC_{0005}$ = 0.091 ppm ($EC_{0005}$ = 0.15 ppm). Adding unexposed workers into the calculation changes the unit risk estimate only slightly. Recognizing the high survival rate of NPC patients, EPA also calculated unit risk by using NPC incidence from the NCI SEER database to replace NPC mortality (that is, replace columns D and I with the SEER NPC incidence data). To be consistent, the calculation would also require the use of cancer-incidence data from the NCI cohort in the Poisson dose-response modeling. However, that was not feasible because cancer-incidence data on the NCI cohort (that is, when new cases were first diagnosed) were not available. Nonetheless, EPA's exercise resulted in new estimates of unit risk that are twice those based on mortality data. EPA correctly pointed out that the correction was attributable to substantial survivorship after NPC onset but was based on the assumption that the exposure-response relationship between formaldehyde exposure and cancer mortality was the same as the relationship between exposure and cancer incidence. That assumption is practical but untestable. EPA also reported unit risk estimates based on Hodgkin lymphoma and leukemia mortality obtained from the extended follow-up of the NCI cohort (Beane-Freeman et al. 2009). The analyses followed the same methods that were used for NPC except that a 2-year lag was used instead of 15 years. The extra risk level at the point of departure was 0.05% for Hodgkin lymphoma mortality but 0.5% for leukemia because of the relatively high leukemia mortality observed in the NCI cohort. Unit risk estimates are summarized below in Table 6-2 for the three cancers, using mortality or incidence, including all person-years vs exposed workers only to demonstrate uncertainties and variability as influenced by these factors.

EPA's unit risk estimate for leukemia is greater than that for NPC or Hodgkin lymphoma and reflects the choice of point of departure and the high background leukemia mortality. Unit risk estimates based on cancer incidence are universally greater than those based on mortality because of the substantial survivorship. Although the range of variation in the unit risk estimates does not incorporate the effect of all sources of variation, EPA's estimation is consistent with the principle of variability and uncertainty analysis in risk assessment.

**TABLE 6-2** Cancer Unit Risk Estimates for Formaldehyde

| Cancer | Data | Person-Years | $EC^a$ (ppm) | $LEC^a$ (95%) (ppm) | Unit Risk$^b$ (ppm$^{-1}$) |
|---|---|---|---|---|---|
| NPC | Mortality | Exposed only | 0.15 | 0.091 | $5.5 \times 10^{-3}$ |
| | | All | 0.15 | 0.093 | $5.4 \times 10^{-3}$ |
| | Incidence | Exposed only | 0.072 | 0.045 | $1.1 \times 10^{-2}$ |
| | | All | 0.074 | 0.046 | $1.1 \times 10^{-2}$ |
| Hodgkin lymphoma | Mortality | Exposed only | 0.155 | 0.088 | $5.7 \times 10^{-3}$ |
| | | All | 0.151 | 0.088 | $5.7 \times 10^{-3}$ |
| | Incidence | Exposed only | 0.053 | 0.030 | $1.7 \times 10^{-2}$ |
| | | All | 0.052 | 0.030 | $1.7 \times 10^{-2}$ |
| Leukemia | Mortality | Exposed only | 0.246 | 0.126 | $4.0 \times 10^{-2}$ |
| | | All | 0.224 | 0.121 | $4.1 \times 10^{-2}$ |
| | Incidence | Exposed only | 0.178 | 0.091 | $5.5 \times 10^{-2}$ |
| | | All | 0.162 | 0.088 | $5.7 \times 10^{-2}$ |
| Total cancer$^c$ | Mortality | All | 0.1 | | $4.5 \times 10^{-2}$ |
| | Incidence | All | 0.1 | | $8.1 \times 10^{-2}$ |

$^a$Extra risk level = 0.0005 for NPC and Hodgkin lymphoma, 0.005 for leukemia.
$^b$Unit risk = extra risk/LEC
$^c$Risk associated with total cancer is based on the sum of estimated extra risk of each cancer at an exposure of 0.1 ppm.
Abbreviations: NPC, nasopharyngeal cancer; EC, effective concentration; and LEC, lower confidence limit on the effective concentration.
Source: EPA 2010a.

EPA further derived an estimate of "total cancer" risk by combining risk of the three cancers. First, EPA estimated the lifetime extra risk of each cancer separately at 0.1 ppm and then added the three estimates and computed the upper confidence limit of the sum. The upper confidence limit is reported as the unit total cancer risk: $4.5 \times 10^{-2}$ and $8.1 \times 10^{-2}$ ppm$^{-1}$ for mortality and incidence, respectively (see Table 6-3). EPA's computation amounts to using the sum of risk of each cancer as a conservative approximation of the risk of any (total) cancer and relies on the assumption that the maximum likelihood estimates of the three cancer risks are independent, an assumption that is convenient but needs justification because the estimates were derived from the same sample of person-years of exposure. A statistically sound alternative would be to consider incidence or mortality of any cancer and then follow the same methods for NPC incidence or mortality. That would be a preferred approach but would require EPA to fit a Poisson regression to the total cancer incidence or mortality.

## Sources of Uncertainty

A unit risk estimate is subject to uncertainty and variability attributable to many sources at various stages of the derivation process. Moreover, it is difficult to determine the degree to which each source affects the overall uncertainty and variation in the final estimate. EPA discussed many potential sources of uncertainty involved in the derivation of the final unit risk estimates. It not only qualitatively identified important sources of uncertainty but quantitatively explored the variability and uncertainty with respect to different cancers, points of departure, all person-years vs only exposed person-years, and mortality vs incidence. It also adjusted for susceptibility in earlier-life exposure. Although EPA did a commendable job in evaluating some of the underlying uncertainties, the committee finds that there is room for further improvement, especially in describing and applying systematic inclusion and exclusion criteria for selecting studies and cancer end points and in using alternative dose-response models.

## Estimating Unit Risks by Using Animal Studies

To validate and supplement the unit risk estimates using human data, EPA reanalyzed the nasal squamous cell carcinoma (SCC) incidence data from two long-term bioassays that used F344 rats (Kerns et al. 1983; Monticello et al. 1996). The two bioassays were combined in EPA's reanalysis to achieve a set of robust dose-response data. The combined dataset has SCC incidences of 0% (n = 341), 0% (n = 107), 0% (n = 353), 0.87% (n = 343), 21.4% (n = 103), and 42% (n = 386) in dose groups of 0, 0.7, 2, 6.01, 9.93, and 14.96 ppm, respectively. EPA conducted a dose-response assessment by using a clonal growth model of the nasal tumor with formaldehyde flux to tissue as the dose metric. The analysis resulted in a unit risk of $1.2 \times 10^{-2}$ ppm$^{-1}$ (extra risk, 0.005) and $2.2 \times 10^{-2}$ ppm$^{-1}$ (extra risk, 0.01) for humans after interspecies scaling. The estimates are relatively consistent with the risk estimates derived from human data from the NCI studies. Moreover, EPA characterized uncertainties attributable to dose-response models (Weibull model with threshold, multistage model for time to tumor, and clonal growth model), extra risk level (1%, 5%, or 10%), and dose metric (flux, DPX). The resulting unit risk estimates are in the range of $1.4 \times 10^{-2}$ to $1.9 \times 10^{-1}$ ppm$^{-1}$. The variation confirms increasing unit risk with increasing extra risk level. Uncertainty remained within less than a factor of 3 between various dose-response models. EPA's efforts to conduct independent dose-response assessment are valuable.

## CONCLUSIONS AND RECOMMENDATIONS

The committee reviewed EPA's approach to derivation of the RfCs and unit risks for formaldehyde as described in the draft IRIS assessment. The committee's general conclusions and recommendations to be considered in revision of the draft assessment are provided below.

- The committee supports EPA's selection of effects on which it based candidate RfCs but does not support the advancement of two studies selected by EPA: Ritchie and Lehnen (1987) and Rumchev et al. (2002). Furthermore, the lack of clear selection criteria, inadequate discussion of some modes of action, little synthesis of responses in animal and human studies, and lack of clear rationales for many conclusions weaken EPA's arguments as presented in the draft IRIS assessment.
- The committee disagrees with EPA's decision not to calculate a candidate RfC for upper respiratory tract pathology. Many well-documented studies have reported the occurrence of upper respiratory tract pathology in laboratory animals, including nonhuman primates, after inhalation exposure to formaldehyde, and the committee recommends that EPA use the animal data to calculate a candidate RfC for this end point.
- The committee found that EPA dismissed the results of the exposure chamber and other nonresidential studies too readily. Although the exposure durations for the chamber studies are short relative to the chronic duration of the RfC, the studies provide complementary information that could be used for deriving a candidate RfC.
- Regarding the uncertainty factor that accounts for variability in response of the human population, the committee suggests application of a value of 3 to calculate the candidate RfCs on the basis of the work of Garrett et al. (1999), Hanrahan et al. (1984), and Liu et al. (1991). Those studies included potentially susceptible populations, so the default value of 10 is not necessary. However, uncertainties remain regarding susceptible populations and factors that affect susceptibility, so a value of 1 is not recommended.
- Regarding the uncertainty factor that accounts for database completeness, the committee suggests that EPA apply its first option as described in the

draft IRIS assessment; that is, apply a value of 1 with the qualification that further research on reproductive, developmental, neurotoxic, and immunotoxic effects would be valuable.

• Overall, the committee found little synthesis of the relationships among the identified noncancer health effects; it appeared that EPA was driven by the need to identify the best study for each health effect rather than trying to integrate all the information. The committee strongly recommends the use of appropriate graphic aids that better display the range of concentrations evaluated in each published study selected for quantitative assessment; the figures may help to identify how findings of studies cluster and especially identify low or high reference values that may be inconsistent with the body of literature. Ultimately, such graphics will improve the ability of the assessment and make a compelling case for the RfC ultimately put forward.

• Regarding calculation of unit risks, the committee agrees that the NCI studies and the findings of the two follow-ups are a reasonable choice because they are the only ones with sufficient exposure and dose-response data for risk estimation. However, the studies are not without their weaknesses, and these need to be clearly articulated in the revised IRIS assessment.

• The committee agrees that EPA's choice of NPC, Hodgkin lymphoma, and leukemia data from the NCI studies to estimate a unit risk is appropriate given that the analysis of Hodgkin lymphoma and leukemia primarily supports the assessment of uncertainty and the magnitude of potential cancer risk. However, the mode of action for formaldehyde-induced Hodgkin lymphoma and leukemia has not been clearly established. Moreover, the highly limited systemic delivery of formaldehyde draws into question the biologic feasibility of causality between formaldehyde exposure and the two cancers. Thus, substantial uncertainties in using Hodgkin lymphoma and leukemia for consensus cancer risk estimation remain.

• Overall, the committee finds EPA's approach to calculating the unit risks reasonable. However, EPA should validate the Poisson dose-response models for NPC, leukemia, and Hodgkin lymphoma mortality with respect to adequacy of model fit, including goodness of fit in the low-dose range, (log) linearity, and absence of interactions of covariates with formaldehyde exposure. Furthermore, EPA is strongly encouraged to conduct alternative dose-response modeling by using Cox regression or alternative nonlinear function forms.

• The draft IRIS assessment does not provide adequate narratives regarding selection of studies and end points for derivation of unit risks. The committee strongly recommends that EPA develop, state, and systematically apply a set of selection criteria for studies and cancer end points.

The committee recognizes that uncertainty and variability remain critical issues as EPA continues to promote quantitative assessment to improve environmental regulation. There are still technical gaps in developing and applying quantitative analysis of uncertainty and variability, especially to incorporate from all sources and at all stages into an overall summary. The NRC Committee to Review EPA's Toxicological Assessment of Tetrachloroethylene (NRC 2010) made several recommendations for advancing methodology and promoting applications. Further research is needed to study various approaches. Small (2008) discussed a probabilistic framework. Given a set of options related to a key assumption (such as mode of action) or a key choice (such as cancer end point), a preference score (or prior probability) may be assigned to each option. The final risk estimate thus also has a weight or probability attached that combines the preference on all options over each assumption or choice. The overarching weight is the result of propagation of uncertainty in each assumption or choice and aggregation of all assumptions over the risk assessment process tree. The collection of final risk estimates for all permissible combinations of assumption and choice forms an empirical distribution. That distribution quantifies the full range of variation and uncertainty in the risk estimate. With the full range of variation of risk estimates and other information on preference of key assumptions and choices, regulatory policy can depend less on a single principal study, a single principal dataset, or a principal end point. The risk-management process may use the distributional properties of the risk estimate to choose a final risk estimate in the context of all feasible assumptions and choices. The committee concludes that further development of systematic approaches to quantifying uncertainty and variation will enable EPA to conduct IRIS assessments in a more transparent and objective fashion.

## REFERENCES

Beane-Freeman, L.E., A. Blair, J.H. Lubin, P.A. Stewart, R.B. Hayes, R.N. Hoover, and M. Hauptmann. 2009. Mortality from lymphohematopoietic malignancies among workers in formaldehyde industries: The National Cancer Institute cohort. J. Natl. Cancer Inst. 101(10):751-761.

Bessac, B.F. and S.E. Jordt. 2008. Breathtaking TRP channels: TRPA1 and TRPV1 in airway chemosensation and reflex control. Physiology (Bethesda). 23:360-370.

Blair, A., P. Stewart, M. O'Berg, W. Gaffey, J. Walrath, J. Ward, R. Bales, S. Kaplan, and D. Cubit. 1986. Mortality among industrial workers exposed to formaldehyde. J. Natl. Cancer Inst. 76(6):1071-1084.

Breslow, N.E., and N.E. Day. 1987. Statistical Methods in Cancer Research, Vol. 2. The Design and Analysis of Cohort Study. IARC Scientific Publications No. 82. Lyon: International Agency for Research on Cancer [online]. Available: http://www.iarc.fr/en/publications/pdfs-online/stat/sp82/SP82.pdf [accessed Jan. 25, 2011].

Caceres, A.I., M. Brackmann, M.D. Elia, B.F. Bessac, D. del Camino, M. D'Amours, J.S. Witek, C.M. Fanger, J.A. Chong, N.J. Hayward, R.J. Homer, L. Cohn, X. Huang, M.M. Moran, and S.E. Jordt. 2009. A sensory neuronal ion channel essential for airway inflammation and hyperreactivity in asthma. Proc. Nat. Acad. Sci. 106(22): 9099-9104.

Callas, P.W., H. Pastides, and D.W. Hosmer. 1998. Empirical comparisons of proportional hazards, poisson, and logistic regression modeling of occupational cohort data. Am. J. Ind. Med. 33(1):33-47.

EPA (U.S. Environmental Protection Agency). 1991. Guidelines for Developmental Toxicity Risk Assessment. EPA/600/FR-91/001. Risk Assessment Forum, U.S. Environmental Protection Agency, Washington, DC. December 1991 [online]. Available: http://iccvam.niehs.nih.gov/SuppDocs/FedDocs/EPA/EPA-devtox.pdf [accessed Feb. 28, 2011].

EPA (U.S. Environmental Protection Agency). 1994. Methods for Derivation of Inhalation Reference Concentrations and Application of Inhalation Dosimetry. EPA/600/B-90/066F. Office of Health and Environmental Assessment, Office of Research and Development, U.S. Environmental Protection Agency, Research Triangle Park, NC [online]. Available: http://www.epa.gov/raf/publications/pdfs/RFCMETHODOLOGY.PDF [accessed Nov. 28, 2010].

EPA (U.S. Environmental Protection Agency). 1998. Guidelines for Neurotoxicity Risk Assessment. EPA/630/R-95/001F. Risk Assessment Forum, U.S. Environmental Protection Agency, Washington, DC. April 1998 [online]. Available: http://www.epa.gov/raf/publications/pdfs/NEUROTOX.PDF [accessed Feb. 28, 2011].

EPA (U.S. Environmental Protection Agency). 1999. Toxicology Data Requirements for Assessing Risks of Pesticide Exposure to Children's Health: Report of the Toxicology Working Group of the 10X Task Force. Draft Report, April 28, 1999. U.S. Environmental Protection Agency [online]. Available: http://www.epa.gov/scipoly/sap/meetings/1999/may/10xtx428.pdf [accessed Jan. 25, 2011].

EPA (U.S. Environmental Protection Agency). 2002. A Review of the Reference Dose and Reference Concentration Processes. External Review Draft. EPA/630/P-02/002A. Reference Dose/Reference Concentration (RfD/RfC) Technical Panel, Risk Assessment Forum, U.S. Environmental Protection Agency, Washington, DC [online]. Available: http://www.epa.gov/raf/publications/pdfs/rfdrfcextrevdrft.pdf [accessed Jan. 6, 2010].

EPA (U.S. Environmental Protection Agency). 2005. Guidelines for Carcinogen Risk Assessment. EPA/630/P-03/001F. Risk Assessment Forum, U.S. Environmental Protection Agency, Washington, DC. March 2005 [online]. Available: http://www.epa.gov/raf/publications/pdfs/CANCER_GUIDELINES_FINAL_3-25-05.PDF [accessed Nov. 24, 2010].

EPA (U.S. Environmental Protection Agency). 2006. A Framework for Assessing Health Risk of Environmental Exposures to Children. EPA/600/R-05/093F. National Center for Environmrntal Assessment, Office of Research and Development, U.S. Environmental Protection Agency, Washington, DC. September 2006 [online]. Available: http://cfpub.epa.gov/ncea/cfm/recordisplay.cfm?deid=158363 [accessed Mar. 10, 2010].

EPA (U.S. Environmental Protection Agency). 2008a. Child-Specific Exposure Factors Handbook (Final Report) 2008. EPA/600/R-06/096F. National Center for Environmental Assessment, Office of Research and Development, U.S. Environmental Protection Agency, Washington, DC [online]. Available: http://cfpub.epa.gov/ncea/cfm/recordisplay.cfm?deid=199243 [accessed Jan. 25, 2011].

EPA (U.S. Environmental Protection Agency). 2008b. Toxicological Review of Tetrachloroethylene (Perchloroethylene) (CAS No. 127-18-4) In Support of Summary Information on the Integrated Risk Information System (IRIS). External Review Draft. EPA/635/R-08/011A. U.S. Environmental Protection Agency, Washington, DC. June 2008 [online]. Available: http://cfpub.epa.gov/ncea/cfm/recordisplay.cfm?deid=192423 [accessed Jan.25, 2011].

EPA (U.S. Environmental Protection Agency). 2009a. Risk Assessment Guidance for Superfund Volume I: Human Health Evaluation Manual (Part F, Supplemental Guidance for

Inhalation Risk Assessment). EPA-540-R-070-002. OSWER 9285.7-82. Office of Superfund Remediation and Technology Innovation, U.S. Environmental Protection Agency, Washington, DC. January 2009 [online]. Available: http://www.epa.gov/osw er/riskassessment/ragsf/pdf/partf_200901_final.pdf [accessed Nov. 29, 2010].

EPA (U.S. Environmental Protection Agency). 2009b. Toxicological Review of Trichloroethylene (CAS No. 79-01-6) In Support of Summary Information on the Integrated Risk Information System (IRIS). External Review Draft. EPA/635/R-09/011A. U.S. Environmental Protection Agency, Washington, DC. October 2009 [online]. Available: http://cfpub.epa.gov/ncea/cfm/recordisplay.cfm?deid=215006 [accessed Nov. 22, 2010].

EPA (U.S. Environmental Protection Agency). 2010a. Glossary, EPA Risk Assessment. U.S. Environmental Protection Agency [online]. Available: http://www.epa.gov/ risk_assessment/glossary.htm#r [accessed Nov. 29, 2010].

EPA (U.S. Environmental Protection Agency). 2010b. Toxicological Review of Formaldehyde (CAS No. 50-00-0) – Inhalation Assessment: In Support of Summary Information on the Integrated Risk Information System (IRIS). External Review Draft. EPA/635/R-10/002A. U.S. Environmental Protection Agency, Washington, DC [online]. Available: http://cfpub.epa.gov/ncea/iris_drafts/recordisplay.cfm?de id=223614 [accessed Nov. 22, 2010].

EPA (U.S. Environmental Protection Agency). 2010c. Integrated Risk information (IRIS). U.S. Environmental Protection Agency [online]. Available: http://www. epa.gov/iris/ [accessed Dec. 8, 2010].

EPA SAB (U.S. Environmental Protection Agency Science Advisory Board). 2011. Review of EPA's Draft Assessment Entitled "Toxicological Review of Trichloroethylene" (October 2009). EPA-SAB-11-02. Science Advisory Board, U.S. Environmental Protection Agency, Washington, DC [online]. Available: http://yose mite.epa.gov/sab/sabproduct.nsf/B73D5D39A8F184BD85257817004A1988/$File/ EPA-SAB-11-002-unsigned.pdf [accessed Jan. 13, 2011].

Evans, J.S., and S.J.S. Baird. 1998. Accounting for missing data in noncancer risk assessment. Hum. Ecol. Risk Assess. 4(2):291-317.

Garcia, C.L., M. Mechilli, L.P. De Santis, A. Schinoppi, K. Katarzyna, and F. Palitti. 2009. Relationship between DNA lesions, DNA repair and chromosomal damage induced by acetaldehyde. Mutat. Res. 662(1-2):3-9.

Garrett, M.H., M.A. Hooper, B.M. Hooper, P.R. Rayment, and M.J. Abramson. 1999. Increased risk of allergy in children due to formaldehyde exposure in homes. Allergy 54(4):330-337 [Erratum-Allergy 54(12):1327].

Gerard, C. 2005. Biomedicine. Asthmatics breathe easier when it's SNO-ing. Science 308(5728):1560-1561.

Ginsberg, G., B.P. Foos, and M.P. Firestone. 2005. Review and analysis of inhalation dosimetry methods for application to children's risk assessment. J. Toxicol. Environ. Health A. 68(8):573-615.

Ginsberg, G., B. Foos, R.B. Dzubow, and M. Firestone. 2010. Options for incorporating children's inhaled dose into human health risk assessment. Inhal. Toxicol. 22(8):627-647.

Hanrahan, L.P., K.A. Dally, H.A. Anderson, M.S. Kanarek, and J. Rankin. 1984. Formaldehyde vapor in mobile homes: A cross sectional survey of concentrations and irritant effects. Am. J. Public Health 74(9):1026-1027.

Hauptmann, M., J.H. Lubin, P.A. Stewart, R.B. Hayes, and A. Blair. 2004. Mortality from solid cancers among workers in formaldehyde industries. Am. J. Epidemiol. 159(12):1117-1130.

Hedberg, J.J., M. Backlund, P. Strömberg, S. Lönn, M.L. Dahl, M. Ingelman-Sundberg, and J.O. Höög. 2001. Functional polymorphism in the alcohol dehydrogenase 3 (ADH3) promoter. Pharmacogenetics. 11(9): 815-824.
Holmström, M., and B. Wilhelmsson. 1988. Respiratory symptoms and pathophysiological effects of occupational exposure to formaldehyde and wood dust. Scand. J. Work Environ. Health 14(5):306-311.
Jensen, D.E., G.K. Belka, and G.C. du Bois. 1998. S-Nitrosoglutathione is a substrate for rat alcohol dehydrogenase class III isoenzyme. Biochem. J. 331(Pt. 2):659-668.
Kerns, W.D., K.L. Pavkov, D.J. Donofrio, E.J. Gralla, J.A. Swenberg. 1983. Carcinogenicity of formaldehyde in rats and mice after long-term inhalation exposure. Cancer Res. 43(9):4382-4392.
Kriebel, D., S.R. Sama, and B. Cocanour. 1993. Reversible pulmonary responses to formaldehyde. A study of clinical anatomy students. Am. Rev. Respir. Dis. 148(6 Pt 1):1509-1515.
Krzyzanowski, M., J.J. Quackenboss, and M.D. Lebowitz. 1990. Chronic respiratory effects of indoor formaldehyde exposure. Environ. Res. 52(2):117-125.
Lee, J.T., and C.Y. Ko. 2005. Has survival improved for nasopharyngeal carcinoma in the United States? Otolaryngol. Head Neck Surg. 132(2):303-308.
Liu, K.S., F.Y. Huang, S.B. Hayward, J. Wesolowski, and K. Sexton. 1991. Irritant effects of formaldehyde exposure in mobile homes. Environ. Health Perspect. 94:91-94.
Lowrance, W.W. 1976. Of Acceptable Risk: Science and the Determination of Safety. Los Altos, CA: W. Kaufmann.
Lyapina, M., G. Zhelezova, E. Petrova, and M. Boev. 2004. Flow cytometric determination of neutrophil burst activity in workers exposed to formaldehyde. Int. Arch. Occup. Environ. Health 77(5):335-340.
Macpherson, L.J., B. Xiao, K.Y. Kwan, M.J. Petrus, A.E. Dubin, S. Hwang, B. Cravatt, D.P. Corey, and A. Patapoutian. 2007. An ion channel essential for sensing chemical damage. J. Neurosci. 27(42):11412-11415.
Malek, F.A., K.U. Möritz, and J. Fanghänel. 2003a. A study on the effect of inhalative formaldehyde exposure on water labyrinth test performance in rats. Ann. Anat. 185(3):277-285.
Marsh, G.M., and A.O. Youk. 2005. Reevaluation of mortality risks for nasopharyngeal cancer in the formaldehyde cohort study of the National Cancer Institute. Regul. Toxicol. Pharmacol. 42(3):275-283.
Marsh, G.M., A.O. Youk, J.M. Buchanich, L.D. Cassidy, L.J. Lucas, N.A. Esmen, and I.M. Gathuru. 2002. Pharyngeal cancer mortality among chemical plant workers exposed to formaldehyde. Toxicol. Ind. Health 18(6):257-268.
Marsh, G.M., A.O. Youk, J.M. Buchanich, S. Erdal, and N.A. Esmen. 2007a. Work in the metal industry and nasopharyngeal cancer mortality among formaldehyde-exposed workers. Regul. Toxicol. Pharmacol. 48(3):308-319.
Marsh, G.M., A.O. Youk, and P. Morfeld. 2007b. Mis-specified and non-robust mortality risk models for nasopharyngeal cancer in the National Cancer Institute formaldehyde worker cohort study. Regul. Toxicol. Pharmacol. 47(1):59-67.
Monticello, T.M., J.A. Swenberg, E.A. Gross, J.R. Leininger, J.S. Kimbell, S. Seilkop, T.B. Starr, J.E. Gibson, and K.T. Morgan. 1996. Correlation of regional and nonlinear formaldehyde-induced nasal cancer with proliferating populations of cells. Cancer Res. 56(5):1012-1022.
National Center for Health Statistics. 2002. United States Life Tables, 1999. National Vital Statistics Reports, vol. 50, no. 6. National Center for Health Statistics

[online]. Available: http://www.cdc.gov/nchs/data/nvsr/nvsr50/nvsr50_06.pdf [accessed Feb. 1, 2011].

NRC (National Research Council). 1993. Pesticides in the Diets of Infants and Children. Washington, DC: National Academy Press.

NRC (National Research Council). 2009. Science and Decisions: Advancing Risk Assessment. Washington, DC: National Academies Press.

NRC (National Research Council). 2010. Review of the Environmental Protection Agency's Draft IRIS Assessment of Tetrachloroethylene. Washington, DC: National Academies Press.

Pinkerton, K.E. and J.P. Joad. 2000. The mammalian respiratory system and critical windows of exposure for children's health. Environ. Health Perspect. 108(Suppl 3):457-462.

Que, L.G., L. Liu, Y. Yan, G.S. Whitehead, S.H. Gavett, D.A. Schwartz, and J.S. Stamler. 2005. Protection from experimental asthma by an endogenous bronchodilator. Science 308(5728):1618-1621.

Rice, D. and S. Barone Jr. 2000. Critical periods of vulnerability for the developing nervous system: evidence from humans and animal models. Environ. Health Perspect. 108(Suppl 3):511-533.

Ritchie, I.M., and R.G. Lehnen. 1987. Formaldehyde-related health complaints of residents living in mobile and conventional homes. Am. J. Public Health 77(3):323-328.

Rowland, A.S., D.D. Baird, C.R. Weinberg, D.L. Shore, C.M. Shy, and A.J. Wilcox. 1992. Reduced fertility among women employed as dental assistants exposed to high levels of nitrous oxide. N. Engl. J. Med. 327(14):993-997.

Rumchev, K.B., J.T. Spickett, M.K. Bulsara, M.R. Phillips, and S.M. Stick. 2002. Domestic exposure to formaldehyde significantly increases the risk of asthma in young children. Eur. Respir. J. 20(2):403-408.

Small, M.J. 2008. Methods for assessing uncertainty in fundamental assumptions and associated models for cancer risk assessment. Risk Anal. 28(5):1289-1308.

Taskinen, H.K., P. Kyyronen, M. Sallmen, S.V. Virtanen, T.A. Liukkonen, O. Huida, M.L. Lindbohm, and A. Anttila. 1999. Reduced fertility among female wood workers exposed to formaldehyde. Am. J. Ind. Med. 36(1):206-212.

Thompson, C.M., and R.C. Grafstrom. 2008. Mechanistic considerations for formaldehyde-induced bronchoconstriction involving S-nitroglutathione reductase. J. Toxicol. Environ. Health A. 71(3):244-248.

Thompson, C.M., B. Sonawane, and R.C. Grafstrom. 2009. The ontogeny, distribution, and regulation of alcohol dehydrogenase 3: Implications for pulmonary physiology. Drug Metab. Dispos. 37(8):1565-1571.

Thompson, C.M., R. Ceder, and R.C. Grafström. 2010. Formaldehyde dehydrogenase: beyond phase I metabolism. Toxicol Lett. 193(1):1-3.

Wu, H., I. Romieu, J.J. Sienra-Monge, B.E. Del Rio-Navarro, D.M. Anderson, C.A. Jenchura, H. Li, M. Ramirez-Aguilar, I. Del Carmen Lara-Sanchez, and S.J. London. 2007. Genetic variation in S-nitrosoglutathione reductase (GSNOR) and childhood asthma. J. Allergy Clin. Immunol. 120(2):322-328.

# 7

# A Roadmap for Revision

In reviewing the draft assessment *Toxicological Review of Formaldehyde-Inhalation Assessment: In Support of Summary Information on the Integrated Risk Information System (IRIS)*, the committee initially evaluated the general methodology (Chapter 2) and then considered the dosimetry and toxicology of formaldehyde (Chapter 3) and the review of the evidence and selection of studies related to noncancer and cancer outcomes (Chapters 4 and 5). Finally, the committee addressed the calculation of the reference concentrations (RfCs) for noncancer effects and the unit risks for cancer and the treatment of uncertainty and variability (Chapter 6). In this chapter, the committee provides general recommendations for changes that are needed to bring the draft to closure. On the basis of "lessons learned" from the formaldehyde assessment, the committee offers some suggestions for improvements in the IRIS development process that might help the Environmental Protection Agency (EPA) if it decides to modify the process. As noted in Chapter 2, the committee distinguishes between the process used to generate the draft IRIS assessment (that is, the development process) and the overall process that includes the multiple layers of review. The committee is focused on the development of the draft IRIS assessment.

## CRITICAL REVISIONS OF THE CURRENT DRAFT IRIS ASSESSMENT OF FORMALDEHYDE

The formaldehyde draft IRIS assessment has been under development for more than a decade (see Chapter 1, Figure 1-3), and its completion is awaited by diverse stakeholders. Here, the committee offers general recommendations—in addition to its specific recommendations in Chapters 3-6—for the revisions that are most critical for bringing the document to closure. Although the committee suggests addressing some of the fundamental aspects of the approach to generating the draft assessment later in this chapter, it is not recommending that the assessment for formaldehyde await the possible development of a revised approach. The following recommendations are viewed as critical overall changes needed to complete the draft IRIS assessment:

- To enhance the clarity of the document, the draft IRIS assessment needs rigorous editing to reduce the volume of text substantially and address redundancy and inconsistency. Long descriptions of particular studies, for example, should be replaced with informative evidence tables. When study details are appropriate, they could be provided in appendixes.
- Chapter 1 needs to be expanded to describe more fully the methods of the assessment, including a description of search strategies used to identify studies with the exclusion and inclusion criteria clearly articulated and a better description of the outcomes of the searches (a model for displaying the results of literature searches is provided later in this chapter) and clear descriptions of the weight-of-evidence approaches used for the various noncancer outcomes. The committee emphasizes that it is not recommending the addition of long descriptions of EPA guidelines to the introduction, but rather clear concise statements of criteria used to exclude, include, and advance studies for derivation of the RfCs and unit risk estimates.
- Standardized evidence tables for all health outcomes need to be developed. If there were appropriate tables, long text descriptions of studies could be moved to an appendix or deleted.
- All critical studies need to be thoroughly evaluated with standardized approaches that are clearly formulated and based on the type of research, for example, observational epidemiologic or animal bioassays. The findings of the reviews might be presented in tables to ensure transparency. The present chapter provides general guidance on approaches to reviewing the critical types of evidence.
- The rationales for the selection of the studies that are advanced for consideration in calculating the RfCs and unit risks need to be expanded. All candidate RfCs should be evaluated together with the aid of graphic displays that incorporate selected information on attributes relevant to the database.
- Strengthened, more integrative, and more transparent discussions of weight of evidence are needed. The discussions would benefit from more rigorous and systematic coverage of the various determinants of weight of evidence, such as consistency.

## FUTURE ASSESSMENTS AND THE IRIS PROCESS

This committee's review of the draft IRIS assessment of formaldehyde identified both specific and general limitations of the document that need to be addressed through revision. The persistence of limitations of the IRIS assessment methods and reports is of concern, particularly in light of the continued evolution of risk-assessment methods and the growing societal and legislative pressure to evaluate many more chemicals in an expedient manner. Multiple groups have recently voiced suggestions for improving the process. The seminal "Red Book," the National Research Council (NRC) report *Risk Assessment in the Federal Government: Managing the Process*, was published in 1983 (NRC 1983). That report provided the still-used four-element framework for risk assessment: hazard identification, dose-response assessment, exposure assessment, and risk characterization. Most recently, in the "Silver Book," *Science and Decisions: Advancing Risk Assessment*, an NRC committee extended the framework of the Red Book in an effort to make risk assessments more useful for decision-making (NRC 2009). Those and other reports have consistently highlighted the necessity for comprehensive assessment of evidence and characterization of uncertainty and variability, and the Silver Book emphasizes assessment of uncertainty and variability appropriate to the decision to be made.

*Science and Decisions: Advancing Risk Assessment* made several recommendations directly relevant to developing IRIS assessments, including the draft formaldehyde assessment. First, it called for the development of guidance related to the handling of uncertainty and variability, that is, clear definitions and methods. Second, it urged a unified dose-response assessment framework for chemicals that would link understanding of disease processes, modes of action, and human heterogeneity among cancer and noncancer outcomes. Thus, it suggested an expansion of cancer dose-response assessments to reflect variability and uncertainty more fully and for noncancer dose-response assessments to reflect analysis of the probability of adverse responses at particular exposures. Although that is an ambitious undertaking, steps toward a unifying framework would benefit future IRIS assessments. Third, the Silver Book recommended that EPA assess its capacity for risk assessment and take steps to ensure that it is able to carry out its challenging risk-assessment agenda. For some IRIS assessments, EPA appears to have difficulty in assembling the needed multidisciplinary teams.

The committee recognizes that EPA has initiated a plan to revise the overall IRIS process and issued a memorandum that provided a brief description of the steps (EPA 2009a). Figure 7-1 illustrates the steps outlined in that memorandum. The committee is concerned that little information is provided on what it sees as the most critical step, that is, completion of a draft IRIS assessment. In the flow diagram, six steps are devoted to the review process, and thus the focus of the revision appears to be on the steps after the assessment has been generated. Although EPA may be revising its approaches for completing the draft assessment (Step 1 in Figure 7-1), the committee could not locate any other information on the revision of the IRIS process. Therefore, the committee offers some suggestions on the development process.

In providing guidance on revisions of the IRIS development process (that is, Step 1 as illustrated in Figure 7-1), the committee begins with a discussion of the current state of science regarding reviews of evidence and cites several examples that provide potential models for IRIS assessments. The

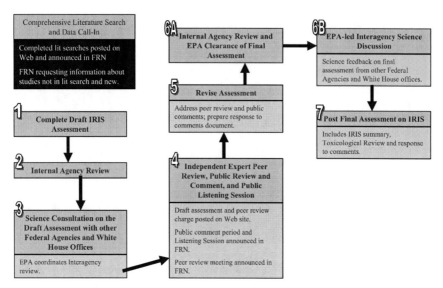

**FIGURE 7-1** New IRIS assessment process. Abbreviations: FRN, Federal Register Notice; IRIS, Integrated Risk Information System; and EPA, Environmental Protection Agency. Source: EPA 2009a.

committee also describes the approach now followed in reviewing and synthesizing evidence related to the National Ambient Air Quality Standards (NAAQSs), a process that has been modified over the last 2 years. It is provided as an informative example of how the agency was able to revise an entrenched process in a relatively short time, not as an example of a specific process that should be adopted for the IRIS process. Finally, the committee offers some suggestions for improving the IRIS development process, providing a "roadmap" of the specific items for consideration.

### An Overview of the Development of the Draft IRIS Assessment

In Chapter 2, the committee provided its own diagram (Figure 2-1) describing the steps used to generate the draft IRIS assessment. For the purpose of offering committee comments on ways to improve those steps, that figure has been expanded to indicate the key outcomes at each step (Figure 7-2). For each of the steps, the figure identifies the key questions addressed in the process. At the broadest level, the steps include systematic review of evidence, hazard identification using a weight-of-evidence approach, and dose-response assessment.

The systematic review process is undertaken to identify all relevant literature on the agent of interest, to evaluate the identified studies, and possibly to

# A Roadmap for Revision

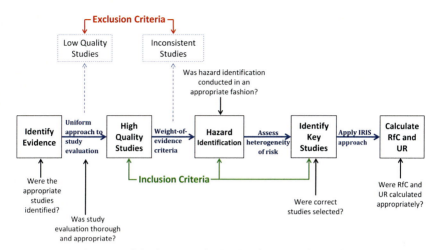

**FIGURE 7-2** Elements of the key steps in the development of a draft IRIS assessment. Abbreviations: IRIS, Integrated Risk Information System; RfC, reference concentration; and UR, unit risk.

provide a qualitative or quantitative synthesis of the literature. Chapter 1 of the draft IRIS assessment of formaldehyde provides a brief general description of the process followed by EPA, including the approach to searching the literature. However, neither Chapter 1 nor other chapters of the draft provide a sufficiently detailed description of the approach taken in evaluating individual studies. In discussing particular epidemiologic studies, a systematic approach to study evaluation is not provided. Consequently, some of the key methodologic points are inconsistently mentioned, such as information bias and confounding.

For hazard identification, the general guidance is also found in Chapter 1 of the draft IRIS assessment. The approach to conducting hazard identification is critical for the integrity of the IRIS process. The various guidelines cited in Chapter 1 provide a general indication of the approach to be taken to hazard identification but do not offer a clear template for carrying it out. For the formaldehyde assessment, hazard identification is particularly challenging because the outcomes include cancer and multiple noncancer outcomes. The various EPA guidelines themselves have not been harmonized, and they provide only general guidance. Ultimately, the quality of the studies reviewed and the strength of evidence provided by the studies for deriving RfCs and unit risks need to be clearly presented. More formulaic approaches are followed for calculation of RfCs and unit risks. The key issue is whether the calculations were conducted appropriately and according to accepted assessment procedures.

## Brief Review of Established Best Practices

The following sections highlight some best practices of current approaches to evidence-based reviews, hazard identification, and dose-response assessment that could provide EPA guidance if it decides to address some of the fundamental issues identified by the committee. The discussion is meant not to be comprehensive or to provide all perspectives on the topics but simply to highlight some important aspects of the approaches. The committee recognizes that some of the concepts and approaches discussed below are elementary and are addressed in some of EPA's guidelines. However, the current state of the formaldehyde draft IRIS assessment suggests that there might be a problem with the practical implementation of the guidelines in completing the IRIS assessments. Therefore, the committee highlights aspects that it finds most critical.

## Current Approaches to Evidence-Based Reviews

Public-health decision-making has a long history of using comprehensive reviews as the foundation for evaluating evidence and selecting policy options. The landmark 1964 report of the U.S. surgeon general on tobacco and disease is exemplary (DHEW 1964). It used a transparent method that involved a critical survey of all relevant literature by a neutral panel of experts and an explicit framework for assessing the strength of evidence for causation that was equivalent to hazard identification (Table 7-1).

The tradition of comprehensive, evidence-based reviews has been continued in the surgeon general's reports. The 2004 surgeon general's report, which marked the 40th anniversary of the first report, highlighted the approach for causal inference used in previous reports and provided an updated and standardized four-level system for describing strength of evidence (DHHS 2004) (Table 7-2).

The same systematic approaches have become fundamental in many fields of clinical medicine and public health. The paradigm of "evidence-based medicine" involves the systematic review of evidence as the basis of guidelines. The international Cochrane Collaboration engages thousands of researchers and clinicians throughout the world to carry out reviews. In the United States, the Agency for Healthcare Research and Quality supports 14 evidence-based practice centers to conduct reviews related to healthcare.

There are also numerous reports from NRC committees and the Institute of Medicine (IOM) that exemplify the use of systematic reviews in evaluating evidence. Examples include reviews of the possible adverse responses associated with Agent Orange, vaccines, asbestos, arsenic in drinking water, and secondhand smoke. A 2008 IOM report, *Improving the Presumptive Disability Decision-Making Process for Veterans*, proposed a comprehensive new scheme for

**TABLE 7-1** Criteria for Determining Causality

| Criterion | Definition |
| --- | --- |
| Consistency | Persistent association among different studies in different populations |
| Strength of association | Magnitude of the association |
| Specificity | Linkage of specific exposure to specific outcome |
| Temporality | Exposure comes before effect |
| Coherence, plausibility, analogy | Coherence of the various lines of evidence with a causal relationship |
| Biologic gradient | Presence of increasing effect with increasing exposure (dose-response relationship) |
| Experiment | Observations from "natural experiments," such as cessation of exposure (for example, quitting smoking) |

Source: DHHS 2004.

**TABLE 7-2** Hierarchy for Classifying Strength of Causal Inferences on the Basis of Available Evidence

A.  Evidence is *sufficient* to infer a causal relationship.

B.  Evidence is *suggestive but not sufficient* to infer a causal relationship.

C.  Evidence is *inadequate* to infer the presence or absence of a causal relationship (evidence that is sparse, of poor quality, or conflicting).

D.  Evidence is *suggestive of no causal relationship*.

Source: DHHS 2004.

evaluating evidence that an exposure sustained in military service had contributed to disease (IOM 2008); the report offers relevant coverage of the practice of causal inference.

This brief and necessarily selective coverage of evidence reviews and evaluations shows that models are available that have proved successful in practice. They have several common elements: transparent and explicitly documented methods, consistent and critical evaluation of all relevant literature, application of a standardized approach for grading the strength of evidence, and clear and consistent summative language. Finally, highlighting features and limitations of the studies for use in quantitative assessments seems especially important for IRIS literature reviews.

A state-of-the-art literature review is essential for ensuring that the process of gathering evidence is comprehensive, transparent, and balanced. The committee suggests that EPA develop a detailed search strategy with search terms related to the specific questions that are addressed by the literature review. The yield of articles from searches can best be displayed graphically, documenting how initial search findings are narrowed to the articles in the final review selection on the basis of inclusion and exclusion criteria. Figure 7-3 provides an example of the selection process in a systematic review of a drug for lung disease. The progression from the initial 3,153 identified articles to the 11 reviewed is transparent. Although this example comes from an epidemiologic meta-analysis, a similar transparent process in which search terms, databases, and resources are listed and study selection is carefully tracked may be useful at all stages of the development of the IRIS assessment.

After studies are identified for review, the next step is to summarize the details and findings in evidence tables. Typically, such tables provide a link to the references, details of the study populations and methods, and key findings. They are prepared in a rigorous fashion with quality-assurance measures, such as using multiple abstractors (at least for a sample) and checking all numbers abstracted. If prepared correctly, the tables eliminate the need for long descriptions of studies and result in shorter text. Some draft IRIS assessments have begun to use a tabular format for systematic and concise presentation of evidence, and the committee encourages EPA to refine and expand that format as it revises the formaldehyde draft IRIS assessment and begins work on others.

The methods and findings of the studies are then evaluated with a standardized approach. Templates are useful for this purpose to ensure uniformity of approach, particularly if multiple reviewers are involved. Such standardized approaches are applied whether the research is epidemiologic (observational), experimental (randomized clinical trials), or toxicologic (animal bioassays). For example, for an observational epidemiologic study, a template for evaluation should consider the following:

- Approach used to identify the study population and the potential for selection bias.
- Study population characteristics and the generalizability of findings to other populations.
- Approach used for exposure assessment and the potential for information bias, whether differential (nonrandom) or nondifferential (random).
- Approach used for outcome identification and any potential bias.
- Appropriateness of analytic methods used.
- Potential for confounding to have influenced the findings.
- Precision of estimates of effect.
- Availability of an exposure metric that is used to model the severity of adverse response associated with a gradient of exposures.

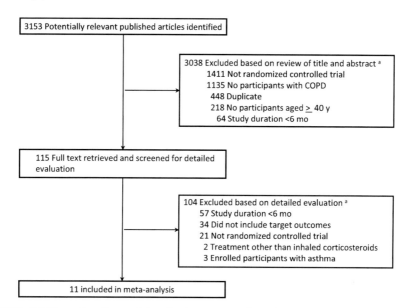

**FIGURE 7-3** Example of an article-selection process. [a]Articles could be excluded for more than one reason; therefore, summed exclusions exceed total. Abbreviation: COPD, chronic obstructive pulmonary disease. Source: Drummond et al. 2008. Reprinted with permission; copyright 2008, American Medical Association.

Similarly, a template for evaluation of a toxicology study in laboratory animals should consider the species and sex of animals studied, dosing information (dose spacing, dose duration, and route of exposure), end points considered, and the relevance of the end points to human end points of concern.

**Current Approaches to Hazard Identification**

Hazard identification involves answering the question, Does the agent cause the adverse effect? (NRC 1983, 2009). Numerous approaches have been used for this purpose, and there is an extensive literature on causal inference, both on its philosophic underpinnings and on methods for evaluating the strength of evidence of causation. All approaches have in common a systematic identification of relevant evidence, criteria for evaluating the strength of evidence, and language for describing the strength of evidence of causation. The topic of causal inference and its role in decision-making was recently covered in the 2008 IOM report on evaluation of the presumptive decision-making process noted above. The 2004 report of the U.S. surgeon general on smoking and health (DHHS 2004) provided an updated review of the methods used in that series of reports.

The review approach for hazard identification embodies the elements described above and uses the criteria for evidence evaluation that have their origins in the 1964 report of the U.S. surgeon general (DHEW 1964) and the writings of Austin Bradford Hill, commonly known as the Hill criteria (see Table 7-1; Hill 1965). The criteria are not rigid and are not applied in a check-list manner; in fact, none is required for inferring a causal relationship, except for temporality inasmuch as exposure to the causal agent must precede the associated effect. The conclusion of causal inference is a clear statement on the strength of evidence of causation. For the purpose of hazard identification, such statements should follow a standardized classification to avoid ambiguity and to ensure comparability among different agents and outcomes.

Beyond the surgeon general's reports used here as an example, there are numerous examples of systematic approaches to hazard identification, including the monographs on carcinogenicity of the International Agency for Research on Cancer and the National Toxicology Program.[1] They have the same elements of systematic gathering and review of all lines of evidence and classification of the strength of evidence in a uniform and hierarchic structure.

**Current Approaches to Dose-Response Assessment**

The topic of dose-response assessment was covered in *Science and Decisions* (NRC 2009), which reviewed the current paradigm and called for a unified framework, bringing commonality to approaches for cancer and noncancer end points. That report also provides guidance on enhancing methods used to characterize uncertainty and variability. The present committee supports those recommendations but offers additional suggestions on the complementary coverage of the use of meta-analysis and pooled analysis in dose-response assessment.

IRIS assessments should address the following critical questions: Which studies should be included for derivation of reference values for noncancer outcomes and unit risks for cancer outcomes? Which dose-response models should be used for deriving those values? The latter question is related to model uncertainty in quantitative risk assessment and is not addressed here in this report. The former question is related to a fundamental issue of filtering the literature to identify the studies that provide the best dose-response information. A related question arises about how to combine information among studies because multiple studies may provide sufficient dose-response data. For this section, the committee assumes that the previously described evidence-based review has identified studies with adequate dose-response information to support some quantification of risk associated with exposure.

As suggested above, it would be unusual for a single study to trump all other studies providing information for setting reference values and unit risks. The combination of the analysis outcomes of different studies falls under the

---

[1]See http://monographs.iarc.fr/index.php and http://ntp.niehs.nih.gov/.

general description of meta-analysis (Normand 1999). The combination and synthesis of results of different studies appears central to an IRIS assessment, but such analyses require careful framing.

Stroup and colleagues (2000) provide a summary of recommendations for reporting meta-analyses of epidemiologic studies. Their proposal includes a table with a proposed check list that has broad categories for reporting, including background (such as problem definition and study population), search strategy (such as searchers, databases, and registries used), methods, results (such as graphic and tabular summaries, study description, and statistical uncertainty), discussion (such as bias and quality of included studies), and conclusion (such as generalization of conclusions and alternative explanations). Their recommendations on methods warrant specific consideration with reference to the development of an IRIS assessment, particularly those on evaluation and assessment of study relevance, rationale for selection and coding of studies, confounding, study quality, heterogeneity, and statistical methods. For the latter, key issues include the selection of models, the clarity with which findings are presented, and the availability of sufficient details to facilitate replication.

In combining study information, it is important that studies provide information on the same quantitative outcome, are conducted under similar conditions, and are of similar quality. If studies are of different quality, this might be addressed by weighting.

The simplest form of combining study information involves the aggregation of p values among a set of independent studies of the same null hypothesis. That simple approach might have appeal for establishing the relationship between some risk factor and an adverse outcome, but it is not useful for establishing exposure levels for a hazard. Thus, effect-size estimation among studies is usually of more interest for risk-estimation purposes and causality assessment. In this situation, a given effect is estimated for each study, and a combined estimate is obtained as a weighted average of study-specific effects in which the weights are inversely related to the precision associated with the estimation of each study-specific effect.

The question is whether EPA should routinely conduct meta-analysis for its IRIS assessments. Implicitly, the development of an IRIS assessment involves many of the steps associated with meta-analysis, including the collection and assessment of background literature. Assuming the availability of independent studies of the same end point and a comprehensive and unbiased inclusion of studies, questions addressed by a meta-analysis may be of great interest. Is there evidence of a homogeneous effect among studies? If not, can one understand the source of heterogeneity? If it is determined that a combined estimate is of interest (for example, an estimate of lifetime cancer risk based on combining study-specific estimates of this risk), a weighted estimate might be derived and reported.

## Case Study: Revision of the Approach to Evidence Review and Risk Assessment for National Ambient Air Quality Standards

Approaches to evidence review and risk assessment vary within EPA. The recently revised approach used for NAAQSs offers an example that is particularly relevant because it represents a major change in an approach taken by one group in the National Center for Environmental Assessment. (EPA 2009b, 2010a,b)

Under Section 109 of the Clean Air Act, EPA is required to consider revisions of the NAAQSs for specified criteria air pollutants—currently particulate matter (PM), ozone, nitrogen dioxide, sulfur dioxide, carbon monoxide, and lead—every 5 years. Through 2009, the process for revision involved the development of two related documents that were both reviewed by the Clean Air Scientific Advisory Committee (CASAC) and made available for public comment. The first, the criteria document, was an encyclopedic compilation, sometimes several thousand pages long, of most scientific publications on the criteria pollutant that had been published since the previous review. Multiple authors contributed to the document, and there was generally little synthesis of the evidence, which was not accomplished in a systematic manner.

The other document was referred to as the staff paper. It was written by a different team in the Office of Air Quality Policy and Standards, and it identified the key scientific advances in the criteria document that were relevant to revising the NAAQSs. In the context of those advances, it offered the array of policy options around retaining or revising the NAAQSs that could be justified by recent research evidence. The linkages between the criteria document and the staff paper were general and not transparent.

The identified limitations of the process led to a proposal for its revision, and it took 2 years to complete the changes in the process. The new process replaces the criteria document with an integrated science assessment and a staff paper that includes a policy assessment. For the one pollutant, PM, that has nearly completed the full sequence, a risk and exposure analysis was also included.

The new documents address limitations of those used previously. The integrated science assessment is an evidence-based review that targets new studies as before. However, review methods are explicitly stated, and studies are reviewed in an informative and purposeful manner rather than in encyclopedic fashion. A main purpose of the integrated science assessment is to assess whether adverse health effects are causally linked to the pollutant under review. The integrated science assessment offers a five-category grading of strength of evidence on each outcome and follows the general weight-of-evidence approaches long used in public health. The intent is to base the risk and exposure analysis on effects for which causality is inferred or those at lower levels if they have particular public-health significance. The risk and exposure analysis brings together the quantitative information on risk and exposure and provides estimates of the current burden of attributable morbidity and mortality and the estimates of avoidable and residual morbidity and mortality under various scenarios of changes in the NAAQS. Standard descriptors for uncertainty are now in place.

The policy assessment develops policy options on the basis of the findings of the integrated science assessment and the risk and exposure analysis. The policy assessment for the PM NAAQS is framed around a series of policy-relevant questions, such as, Does the available scientific evidence, as reflected in the integrated science assessment, support or call into question the adequacy of the protection afforded by the current 24-hr $PM_{10}$ standard against effects associated with exposures to thoracic coarse particles? Evidence-based answers to the questions are provided with a reasonably standardized terminology for uncertainty.

For the most recent reassessment of the PM NAAQS, EPA staff and CASAC found the process to be effective; it led to greater transparency in evidence review and development of policy options than the prior process (Samet 2010). As noted above, the present committee sees the revision of the NAAQS review process as a useful example of how the agency was able to revise an entrenched process in a relatively short time.

## Reframing the Development of the IRIS Assessment

The committee was given the broad charge of reviewing the formaldehyde draft IRIS assessment and also asked to consider some specific questions. In addressing those questions, the committee found, as documented in Chapter 2, that some problems with the draft arose because of the processes and methods used to develop the assessment. Other committees have noted some of the same problems. Accordingly, the committee suggests here steps that EPA could take to improve IRIS assessment through the implementation of methods that would better reflect current practices. The committee offers a roadmap for changes in the development process if EPA concludes that such changes are needed. The term *roadmap* is used because the topics that need to be addressed are set out, but detailed guidance is not provided because that is seen as beyond the committee's charge. The committee's discussion of a reframing of the IRIS development process is based on its generic representation provided in Figure 7-2. The committee recognizes that the changes suggested would involve a multiyear process and extensive effort by the staff of the National Center for Environmental Assessment and input and review by the EPA Science Advisory Board and others. The recent revision of the NAAQS review process provides an example of an overhauling of an EPA evidence-review and risk-assessment process that took about 2 years.

In the judgment of the present and past committees, consideration needs to be given to how each step of the process could be improved and gains made in transparency and efficiency. Models for conducting IRIS reviews more effectively and efficiently are available. For each of the various components (Figure 7-2), methods have been developed, and there are exemplary approaches in assessments carried out elsewhere in EPA and by other organizations. In addition, there are relevant examples of evidence-based algorithms that EPA could draw on. Guidelines and protocols for the conduct of evidence-based reviews are available, as are guidelines for inference as to the strength of evidence of association and causation. Thus, EPA may be able to make changes in the assessment process relatively quickly by drawing on appropriate experts and selecting and adapting existing approaches.

One major, overarching issue is the use of weight of evidence in hazard identification. The committee recognizes that the terminology is embedded in various EPA guidelines (see Appendix B) and has proved useful. The determination of weight of evidence relies heavily on expert judgment. As called for by others, EPA might direct effort at better understanding how weight-of-evidence determinations are made with a goal of improving the process (White et al. 2009).

The committee highlights below what it considers critical for the development of a scientifically sound IRIS assessment. Although many elements are basic and have been addressed in the numerous EPA guidelines, implementation does not appear to be systematic or uniform in the development of the IRIS assessments.

**General Guidance for the Overall Process**

- Elaborate an overall, documented, and quality-controlled process for IRIS assessments.
- Ensure standardization of review and evaluation approaches among contributors and teams of contributors; for example, include standard approaches for reviews of various types of studies to ensure uniformity.
- Assess disciplinary structure of teams needed to conduct the assessments.

**Evidence Identification: Literature Collection and Collation Phase**

- Select outcomes on the basis of available evidence and understanding of mode of action.
- Establish standard protocols for evidence identification.
- Develop a template for description of the search approach.
- Use a database, such as the Health and Environmental Research Online (HERO) database, to capture study information and relevant quantitative data.

## Evidence Evaluation: Hazard Identification and Dose-Response Modeling

- Standardize the presentation of reviewed studies in tabular or graphic form to capture the key dimensions of study characteristics, weight of evidence, and utility as a basis for deriving reference values and unit risks.
- Develop templates for evidence tables, forest plots, or other displays.
- Establish protocols for review of major types of studies, such as epidemiologic and bioassay.

## Weight-of-Evidence Evaluation: Synthesis of Evidence for Hazard Identification

- Review use of existing weight-of-evidence guidelines.
- Standardize approach to using weight-of-evidence guidelines.
- Conduct agency workshops on approaches to implementing weight-of-evidence guidelines.
- Develop uniform language to describe strength of evidence on noncancer effects.
- Expand and harmonize the approach for characterizing uncertainty and variability.
- To the extent possible, unify consideration of outcomes around common modes of action rather than considering multiple outcomes separately.

## Selection of Studies for Derivation of Reference Values and Unit Risks

- Establish clear guidelines for study selection.
  - Balance strengths and weaknesses.
  - Weigh human vs experimental evidence.
  - Determine whether combining estimates among studies is warranted.

## Calculation of Reference Values and Unit Risks

- Describe and justify assumptions and models used. This step includes review of dosimetry models and the implications of the models for uncertainty factors; determination of appropriate points of departure (such as benchmark dose, no-observed-adverse-effect level, and lowest observed-adverse-effect level), and assessment of the analyses that underlie the points of departure.
- Provide explanation of the risk-estimation modeling processes (for example, a statistical or biologic model fit to the data) that are used to develop a unit risk estimate.

• Assess the sensitivity of derived estimates to model assumptions and end points selected. This step should include appropriate tabular and graphic displays to illustrate the range of the estimates and the effect of uncertainty factors on the estimates.

• Provide adequate documentation for conclusions and estimation of reference values and unit risks. As noted by the committee throughout the present report, sufficient support for conclusions in the formaldehyde draft IRIS assessment is often lacking. Given that the development of specific IRIS assessments and their conclusions are of interest to many stakeholders, it is important that they provide sufficient references and supporting documentation for their conclusions. Detailed appendixes, which might be made available only electronically, should be provided when appropriate.

## REFERENCES

DHEW (U.S. Deaprtment of Health Education and Welfare). 1964. Smoking and Health. Report of the Advisory Committee to the Surgeon General. Public Health Service Publication No. 1103. Washington, DC: U.S. Government Printing Office [online]. Available: http://profiles.nlm.nih.gov/NN/B/B/M/Q/_/nnbbmq.pdf [accessed Feb. 1, 2011].

DHHS (U.S. Department of Health and Human Services). 2004. The Health Consequences of Smoking: A Report of the Surgeon General. U.S. Department of Health and Human Services, Centers for Disease Control and Prevention, National Center for Chronic Disease Prevention and Health Promotion, Office on Smoking and Health, Atlanta, GA [online]. Available: http://www.cdc.gov/tobacco/data_statistics/sgr/2004/complete_report/index.htm [accessed Nov. 22, 2010].

Drummond, M.B., E.C. Dasenbrook, M.W. Pitz, D.J. Murphy, and E. Fan. 2008. Inhaled corticosteroids in patients with stable chronic obstructive pulmonary disease: A systematic review and meta-analysis. JAMA. 300(20):2407-2416.

EPA (U.S. Environmental Protection Agency). 2009a. New Process for Development of Integrated Risk Information System Health Assessments. Memorandum to Assistant Administrators, General Counsel, Inspector General, Chief Financial Officer, Chief of Staff, Associate Administrators, and Regional Administrators, from Lisa P. Jackson, the Administrator, U.S. Environmental Protection Agency, Washington, DC. May 21, 2009 [online]. Available: http://www.epa.gov/iris/pdfs/IRIS_PROCESS_MEMO.5.21.09.PDF [accessed Nov. 23, 2010].

EPA (U.S. Environmental Protection Agency). 2009b. Integrated Science Assessment for Particulate Matter (Final Report). EPA/600/R-08/139F. National Center for Environmental Assessment-RTP Division, Office of Research and Development, U.S. Environmental Protection Agency, Research Triangle Park, NC. December 2009 [online]. Available: http://cfpub.epa.gov/ncea/cfm/recordisplay.cfm?deid=216546 [accessed March 2, 2011].

EPA (U.S. Environmental Protection Agency). 2010a. Quantitative Health Risk Assessment for Particulate Matter (Final Report). EPA-452/R-10-005. Office of Air Quality Planning and Standards, Office of Air and Radiation, U.S. Environmental Protection Agency, Research Triangle Park, NC. June 2010 [online]. Available: http://www.epa.gov/ttn/naaqs/standards/pm/data/PM_RA_FINAL_June_2010.pdf [accessed March 2, 2011].

EPA (U.S. Environmental Protection Agency). 2010b. Policy Assessment for the Review of the Particulate Matter National Ambient Air Quality Standards (Second External Review Draft). EPA 452/P-10-007. Office of Air Quality Planning and Standards, Office of Air and Radiation, U.S. Environmental Protection Agency, Research Triangle Park, NC. June 2010 [online]. Available: http://www.epa.gov/ttnnaaqs/standards/pm/data/20100630seconddraftpmpa.pdf [accessed March 2, 2011].

Hill, A.B. 1965. The environment and disease: Association or causation? Proc. R. Soc. Med. 58:295-300

IOM (Institute of Medicine). 2008. Improving the Presumptive Disability Decision-Making Process for Veterans. Washington, DC: National Academies Press.

NRC (National Research Council). 1983. Risk Assessment in the Federal Government: Managing the Process. Washington, DC: National Academy Press.

NRC (National Research Council). 2009. Science and Decisions: Advancing Risk Assessment. Washington, DC: National Academies Press.

Normand, S.L. 1999. Meta-analysis: Formulating, evaluating, combining, and reporting. Stat. Med. 18(3): 321-359.

Samet, J.M. 2010. CASAC Review of Policy Assessment for the Review of the PM NAAQS - Second External Review Draft (June 2010). EPA-CASAC-10-015. Letter to Lisa P. Jackson, Administrator, from Jonathan M. Samet, Clean Air Scientific Advisory Committee, Office of Administrator, Science Advisory Board, U.S. Environmental Protection Agency, Washington, DC. September 10, 2010 [online]. Available: http://yosemite.epa.gov/sab/sabproduct.nsf/CCF9F4C0500C500F8525779D0073C593/$File/EPA-CASAC-10-015-unsigned.pdf [accessed Nov. 22, 2010].

Stroup, D.F., J.A. Berlin, S.C. Morton, I. Olkin, G.D. Williamson, D. Rennie, D. Moher, B.J. Becker, T.A. Sipe, and S.B. Thacker. 2000. Meta-analysis of observational studies in epidemiology: A proposal for reporting. Meta-analysis Of Observational Studies in Epidemiology (MOOSE) group. JAMA. 283(15):2008-2012.

White, R.H., I. Cote, L. Zeise, M. Fox, F. Dominici, T.A. Burke, P.D. White, D.B. Hattis, and J.M. Samet. 2009. State-of-the-science workshop report: Issues and approaches in low-dose-response extrapolation for environmental health risk assessment. Environ. Health Perspect. 117(2):283-287.

# Appendix A

# Biographic Information on the Committee to Review EPA'S Draft IRIS Assessment of Formaldehyde

**Jonathan M. Samet** *(Chair)* is a pulmonary physician and epidemiologist. He is professor and Flora L. Thornton Chair of the Department of Preventive Medicine at the Keck School of Medicine of the University of Southern California (USC) and director of the USC Institute for Global Health. Dr. Samet's research has focused on the health risks associated with inhaled pollutants. He has served on numerous committees concerned with public health: the U.S. Environmental Protection Agency Science Advisory Board; committees of the National Research Council (NRC), including chairing Committee on Health Risks of Exposure to Radon (BEIR VI), the Committee on Research Priorities for Airborne Particulate Matter, and the Board on Environmental Studies and Toxicology; and committees of the Institute of Medicine (IOM). He is a member of the IOM and is chair of the NRC Committee to Develop a Research Strategy for Environmental, Health, and Safety Aspects of Engineered Nanomaterials and a member of the National Academies Committee on Science, Technology, and Law. Dr. Samet received his MD from the University of Rochester, School of Medicine and Dentistry.

**Andrew F. Olshan** *(Vice-Chair)* is professor and chair of the Department of Epidemiology of the University of North Carolina Gillings School of Global Public Health. His research interests are the etiology of birth defects and cancer in children and adults. Recent work has focused on the role of environmental exposures, genetic factors, and adverse health effects in children and adults; risk factors for childhood tumors and neuroblastoma; and the effects of drinking-water disinfection byproducts on male reproductive health. He has served on several National Academies committees, most recently the National Research

Council Committee on Contaminated Drinking Water at Camp Lejeune and the Committee to Review the Evidence Regarding the Link between Exposure to Agent Orange and Diabetes. Dr. Olshan received his PhD in epidemiology from the University of Washington.

**A. John Bailer** is distinguished professor and chair in the Department of Statistics of Miami University in Oxford, Ohio. He is also a research fellow in the university's Scripps Gerontology Center and an affiliate member of the Department of Zoology, the Department of Sociology and Gerontology, and the Institute of Environmental Sciences at Miami University. His research interests include the design and analysis of environmental and occupational health studies and quantitative risk estimation. He has served on several National Research Council Committees, including the Committee on Improving Risk Analysis Approaches Used by the U.S. EPA, the Committee on Spacecraft Exposure Guidelines, the Committee to Review the OMB Risk Assessment Bulletin, and the Committee on Toxicologic Assessment of Low-Level Exposures to Chemical Warfare Agents. He also has served as a member of the Report on Carcinogens Subcommittee and the Technical Reports Review Subcommittee of the Board of Scientific Counselors of the National Toxicology Program. Dr. Bailer received his PhD in biostatistics from the University of North Carolina at Chapel Hill.

**Sandra J.S. Baird** is an environmental analyst with the Massachusetts Department of Environmental Protection Office of Research and Standards. She supports the air toxics and drinking-water programs through the development of cancer and noncancer toxicity values, evaluation of the implications of new toxicologic information and guidance, evaluation of site-specific toxicity and exposure assessment issues, and development of guidance in support of risk-based decision-making. Her research interests include probabilistic characterization of uncertainty in toxicity values for use in risk assessment and mixtures risk assessment. Dr. Baird received her PhD in toxicology from the University of Rochester School of Medicine and Dentistry.

**Harvey Checkoway** is a professor in the Department of Environmental and Occupational Health Sciences and the Department of Epidemiology at the University of Washington School of Public Health and Community Medicine. His expertise is in occupational and environmental determinants of chronic diseases. Research projects for which Dr. Checkoway has been the principal investigator include epidemiologic studies of cancer mortality in nuclear workers, of cancer mortality in phosphate-industry workers, of silicosis and lung cancer in silica-exposed diatomaceous-earth industry workers, of lung cancer in chromate-exposed aerospace workers, of reproductive hazards in lead-smelter workers, of cancer risks and parkinsonism in textile workers, and of environmental and genetic risk factors for Parkinson disease. Dr. Checkoway received his MPH from Yale University and his PhD in epidemiology from the University of North Carolina at Chapel Hill.

**Richard A. Corley** is laboratory fellow in the biologic monitoring and biologic modeling group at the Pacific Northwest National Laboratory operated by Battelle for the U.S. Department of Energy. Dr. Corley specializes in the development of physiologically based pharmacokinetic models, real-time breath analysis, dermal and inhalation bioavailability, and the development of three-dimensional computational fluid-dynamic models of the respiratory system. He has published numerous peer-reviewed papers on oral, dermal, and inhalation toxicology; on modes of action of a variety of industrial and consumer chemicals; and on pharmacokinetic modeling and its applications in human health risk assessment. Dr. Corley served on the National Research Council Committee to Assess the Health Implications of Perchlorate Ingestion and Standing Committee on Risk Analysis Issues and Reviews. He received a PhD in environmental toxicology from the University of Illinois at Urbana-Champaign.

**David C. Dorman** is associate dean for research and graduate studies in the College of Veterinary Medicine of North Carolina State University. The primary objective of his research is to provide a refined understanding of chemically induced neurotoxicity in laboratory animals that will lead to improved assessment of potential neurotoxicity in humans. Dr. Dorman's research interests include neurotoxicology, nasal toxicology, and pharmacokinetics. He served as a member of the National Research Council Committee on Animal Models for Testing Interventions Against Aerosolized Bioterrorism Agents and as member and chair of two Committees on Emergency and Continuous Exposure Guidance Levels for Selected Submarine Contaminants. He received his DVM from Colorado State University. He completed a combined PhD and residency program in toxicology at the University of Illinois at Urbana-Champaign and is a diplomate of the American Board of Veterinary Toxicology and the American Board of Toxicology.

**Charles H. Hobbs** is a senior scientist emeritus at Lovelace Respiratory Research Institute (LRRI) and member of the board of directors of Lovelace Biomedical and Environmental Research Institute. He was formerly director of toxicology at LRRI. His research interests centered on the long-term biologic effects of inhaled materials and the mechanisms by which they occur. His research has ranged from physical and chemical characterization of airborne toxicants to in vitro mechanistic and toxicologic studies in laboratory animals. Dr. Hobbs is an associate of the National Academies and has served on several committees of the National Research Council, including service as chair of the Committee on Animal Models for Testing Interventions Against Aerosolized Bioterrorism Agents, the Committee on Submarine Escape Action Levels, and the Committee on Beryllium Alloy Exposures, and he is currently a member of the Committee on Biodefense at the U.S. Department of Defense. Dr. Hobbs earned a DVM from Colorado State University.

**Michael D. Laiosa** is an assistant professor at the University of Wisconsin–Milwaukee. He had been a research assistant professor of environmental medicine at the University of Rochester. Dr. Laiosa's research interests are focused on how environmental factors influence immunologically based human diseases, such as leukemia and autoimmunity. His specific interests involve identifying developmental and early-life environmental factors that influence cancer risks and autoimmune pathogenesis; identifying prenatal and early postnatal chemopreventive agents that may reduce risk of such diseases as leukemia, atopy, and autoimmune disease; and determining the effect of early-life exposures to chemical mixtures on long-term immunologic health. Dr. Laiosa earned a PhD in microbiology and immunology from the State University of New York Upstate Medical University.

**Ivan Rusyn** is professor in the Department of Environmental Sciences and Engineering in the School of Public Health of the University of North Carolina (UNC) at Chapel Hill. He directs the Laboratory of Environmental Genomics and the Carolina Center for Computational Toxicology in the Gillings School of Global Public Health of UNC. He also serves as associate director of the Curriculum in Toxicology and is a member of the Lineberger Comprehensive Cancer Center, the Center for Environmental Health and Susceptibility, the Bowles Center for Alcohol Studies, and the Carolina Center for Genome Sciences. Dr. Rusyn's laboratory focuses on the mechanisms of action of environmental toxicants and the genetic determinants of susceptibility to toxicant-induced injury. He is a member of the National Research Council Committee on Use of Emerging Science for Environmental Health Decisions and was previously a member of the Committee on Tetrachlorethylene. Dr. Rusyn received his MD from Ukrainian State Medical University in Kiev and his PhD in toxicology from UNC at Chapel Hill.

**Mary Alice Smith** is an associate professor and graduate coordinator of environmental health science at the University of Georgia. Her research interests are developmental toxicology and risk assessment. Her research focuses on the effects of toxicants on reproduction and development, environmental and microbial risk-assessment methodology, and the effects of pathogens on pregnancy and development. She teaches courses in toxicology, developmental and reproductive toxicology, and risk assessment. Dr. Smith is member of the Teratology Society, the Society of Toxicology, and the International Association for Food Protection, and she is co-director of the Academy of the Environment at the University of Georgia. She has served on grant-review panels for the National Institutes of Health, the Food and Drug Administration, the Centers for Disease Control and Prevention, and the Agency for Toxic Substances and Disease Registry. She has also served as a member of expert panels for the International Life Sciences—North America and the Food and Drug Administration, and she is a member of the editorial board of *Reproductive Toxicology*. Dr. Smith earned a PhD in toxicology and pharmacology from the University of Arkansas for Medical Sciences.

**Leslie T. Stayner** is a professor of epidemiology at the University of Illinois at Chicago. His research interests are occupational and environmental epidemiology, epidemiologic methods, risk assessment, and cancer and other chronic diseases. Dr. Stayner is a member of the Society for Epidemiologic Research, the American Public Health Association, and the International Commission on Occupational Health, and he is a fellow of the American College of Epidemiology and the Institute of Medicine of Chicago. He has served on the editorial boards of several journals, including service as contributing editor of the *American Journal of Industrial Medicine* and editorial consultant for the *American Journal of Epidemiology*. He has also served on the Institute of Medicine Committee on Making Best Use of the Agent Orange Exposure Reconstruction Model and the National Research Council Committee on Human Health Risks of Trichloroethylene. Dr. Stayner earned a PhD in epidemiology from the University of North Carolina at Chapel Hill.

**Helen H. Suh** is the program area director for environmental health at the National Opinion Research Center at the University of Chicago. Until recently, Dr. Suh was on the faculty of Harvard University in the department of environmental health. Her research examines critical questions about the public-health consequences of air pollution through interdisciplinary, biologically relevant exposure-assessment research. Specifically, her research combines traditional measurement methods with novel exposure risk-assessment methods. Her work has been published in several journals. Dr. Suh has served on the National Research Council Committee on Estimating Mortality Risk Reduction Benefits from Decreasing Tropospheric Ozone Exposure and is a member of the U.S. Environmental Protection Agency's Clean Air Scientific Advisory Committee. She received her Sc.D. from Harvard University.

**Yiliang Zhu** is professor in the Department of Epidemiology and Biostatistics of the University of South Florida College of Public Health. He is also director of the college's Center for Collaborative Research. His current research is focused on quantitative methods in health risk assessment, including physiologically based pharmacokinetic models, dose-response modeling, benchmark dose methods, and uncertainty quantification. He also conducts research in disease surveillance, health-outcome evaluation, and health-care access and use in developing countries. Dr. Zhu has served as a member of the National Research Council Committee on EPA's Exposure and Human Health Reassessment of TCDD and Related Compounds and the Committee on Tetrachloroethylene. He received his Ph.D. in statistics from the University of Toronto.

**Patrick A. Zweidler-Mckay** is an assistant professor at the University of Texas M. D. Anderson Cancer Center and a member of the faculty of the University of Texas Graduate School of Biomedical Sciences. His research is directed at understanding the critical genetic events that lead to the development of leukemia and to the discovery of novel therapeutic approaches through molecular strategies. Clinically, Dr. Zweidler-McKay specializes in treating children who have particularly difficult or relapsed forms of leukemia and lymphoma. He is certified by the American Board of Pediatrics and is a member of the American Society of Hematology, the American Society of Pediatric Hematology and Oncology, the Harris County Medical Society, and the Texas Medical Society. Dr. Zweidler-McKay earned a PhD in molecular biology and genetics and an MD from Temple University.

# Appendix B

# Weight-of-Evidence Descriptions from U.S. Environmental Protection Agency Guidelines

The text in this appendix was excerpted directly from the indicated guidelines of the U.S. Environmental Protection Agency (EPA).

### GUIDELINES FOR MUTAGENICITY RISK ASSESSMENT

The evidence for a chemical's ability to produce mutations and to interact with the germinal target is integrated into a weight-of-evidence judgment that the agent may pose a hazard as a potential human germ-cell mutagen. All information bearing on the subject, whether indicative of potential concern or not, must be evaluated. Whatever evidence may exist from humans must also be factored into the assessment.

All germ-cell stages are important in evaluating chemicals because some chemicals have been shown to be positive in postgonial stages but not in gonia (Russell et al., 1984). When human exposures occur, effects on postgonial stages should be weighted by the relative sensitivity and the duration of the stages. Chemicals may show positive effects for some endpoints and in some test systems, but negative responses in others. Each review must take into account the limitations in the testing and in the types of responses that may exist.

To provide guidance as to the categorization of the weight of evidence, a classification scheme is presented to illustrate, in a simplified sense, the strength of the information bearing on the potential for human germ-cell mutagenicity. It is not possible to illustrate all potential combinations of evidence, and considerable judgment must be exercised in reaching conclusions. In addition, certain responses in tests that do not measure direct mutagenic end points (e.g., SCE induction in mammalian germ cells) may provide a basis for raising the weight of evidence from one category to another. The categories are presented in decreasing order of strength of evidence.

1. Positive data derived from human germ-cell mutagenicity studies, when available, will constitute the highest level of evidence for human mutagenicity.
2. Valid positive results from studies on heritable mutational events (of any kind) in mammalian germ cells.
3. Valid positive results from mammalian germ-cell chromosome aberration studies that do not include an intergeneration test.
4. Sufficient evidence for a chemical's interaction with mammalian germ cells, together with valid positive mutagenicity test results from two assay systems, at least one of which is mammalian (in vitro or in vivo). The positive results may both be for gene mutations or both for chromosome aberrations; if one is for gene mutations and the other for chromosome aberrations, both must be from mammalian systems.
5. Suggestive evidence for a chemical's interaction with mammalian germ cells, together with valid positive mutagenicity evidence from two assay systems as described under 4, above. Alternatively, positive mutagenicity evidence of less strength than defined under 4, above, when combined with sufficient evidence for a chemical's interaction with mammalian germ cells.
6. Positive mutagenicity test results of less strength than defined under 4, combined with suggestive evidence for a chemical's interaction with mammalian germ cells.
7. Although definitive proof of nonmutagenicity is not possible, a chemical could be classified operationally as a nonmutagen for human germ cells if it gives valid negative test results for all endpoints of concern.
8. Inadequate evidence bearing on either mutagenicity or chemical interaction with mammalian germ cells (EPA 1986, Pp 9-10).

## METHODS FOR DERIVATION OF INHALATION REFERENCE CONCENTRATIONS AND APPLICATION OF INHALATION DOSIMETRY

The culmination of the hazard identification phase of any risk assessment involves integrating a diverse data collection into a cohesive, biologically plausible toxicity "picture"; that is, to develop the weight of evidence that the chemical poses a hazard to humans. The salient points from each of the laboratory animal and human studies in the entire data base should be summarized as should the analysis devoted to examining the variation or consistency among factors (usually related to the mechanism of action), in order to establish the likely outcome for exposure to this chemical. From this analysis, an appropriate animal model or additional factors pertinent to human extrapolation may be identified.

The utility of a given study is often related to the nature and quality of the other available data. For example, clinical pharmacokinetic studies may validate that the target organ or disease in laboratory animals is likely to be the same effect observed in the exposed human population. However, if a cohort study describing the nature of the dose-response relationship were available, the clinical description would rarely give additional information. An apparent conflict may arise in the analysis when an association is observed in toxicologic but not epidemiologic data, or vice versa. The analysis then should focus on reasons for the apparent difference in order to resolve the discrepancy. For example, the epidemiologic data may have contained other exposures not accounted for, or the laboratory animal species tested may have been inappropriate for the mechanism of action. A framework for approaching data summary is provided in Table 2-6. Table 2-7 provides the specific uses of various types of human data in such an approach. These guidelines have evolved from criteria used to establish causal significance, such as those developed by the American Thoracic Society (1985) to assess the causal significance of an air toxicant and a health effect. The criteria for establishing causal significance can be found in Appendix C. In general, the following factors enhance the weight of evidence on a chemical:

- Clear evidence of a dose-response relationship;
- Similar effects across sex, strain, species, exposure routes, or in multiple experiments;
- Biologically plausible relationship between metabolism data, the postulated mechanism of action, and the effect of concern;
- Similar toxicity exhibited by structurally related compounds;
- Some correlation between the observed chemical toxicity and human evidence.

The greater the weight of evidence, the greater the confidence in the conclusion derived. Developing improved weight-of-evidence schemes for various noncancer health effect categories has been the focus of efforts by the Agency to improve health risk assessment methodologies (Perlin and McCormack, 1988).

Another difficulty encountered in this summarizing process is that certain studies may produce apparently positive or negative results, yet may be flawed. The flaws may have arisen from inappropriate design or execution in performance (e.g., lack of statistical power or adjustment of dosage during the course of the study to avoid undesirable toxic effects). The treatment of flawed results is critical; although there is something to be learned from every study, the extent that a study should be used is dependent on the nature of the flaw (Society of Toxicology, 1982). A flawed negative study could only provide a false sense of security, whereas a flawed positive study may contribute to some limited understanding. Although there is no substitute for good science, grey areas such as this are ultimately a matter of scientific judgment. The risk assessor will have to decide what is and is not useful within the framework outlined earlier.

Studies meeting the criteria detailed in Sections 2.1.1 and 2.1.2 (epidemiologic, nonepidemiologic data), and experimental studies on laboratory animals that fit into this weight-of- evidence framework are used in the quantitative dose-response assessment discussed in Chapter 4 (EPA 1994, Pp 2-42 to 2-46).

## GUIDELINES FOR DEVELOPMENTAL TOXICITY RISK ASSESSMENT

The 1989 Proposed Amendments described important considerations in determining the relative weight of various kinds of data in estimating the risk of developmental toxicity in humans. The intent of the proposed weight-of-evidence (WOE) scheme was that it not be used in isolation, but be used as the first step in the risk assessment process, to be integrated with dose-response information and the exposure assessment.

The WOE scheme was the subject of a considerable number of public comments, and was one of the major concerns of the SAB. The concern of public commentors was that the reference to human developmental toxicity in this scheme suggested that a chemical could be prematurely designated, and perhaps labeled, as causing developmental toxicity in humans prior to the completion of the risk assessment process. The SAB suggested that the intended use of this scheme was not consistent with the use of the term "weight of evidence" in other contexts, since WOE is usually thought of as an evaluation of the total composite of information available to make a judgment about risk. In addition, the SAB Committee proposed that the Agency consider development of a more conceptual approach using decision analytical techniques to predict the relationships among various outcomes.

In the final Guidelines, the terminology used in the WOE scheme has been completely changed and retitled "Characterization of the Health-Related Database." The intended purpose of the scheme is to provide a framework and criteria for making a decision on whether or not sufficient data are available to conduct a risk assessment. This decision is based on the available data, whether animal or human, and does not necessarily imply human hazard. This decision process is part of, but not the complete, WOE evaluation, which also takes into account the RfDDT or RfCDT and the human exposure information, culminating in risk characterization.

The final Guidelines also place strong emphasis on the integration of the dose-response evaluation with hazard information in characterizing the sufficiency of the health-related database. In line with this approach, the Guidelines have been reorganized to combine hazard identification and dose-response evaluation. Finally, the SAB comments on developing a conceptual matrix provide an interesting challenge, but current data indicate that the relationships among endpoints of developmental toxicity are not consistent across chemicals or species. The Agency is currently supporting modeling efforts to further explore the relationship among various development toxicity endpoints and the development of biologically based dose-response models that consider multiple effects (EPA 1991, Pp 69-70).

## A REVIEW OF THE REFERENCE DOSE AND REFERENCE CONCENTRATION PROCESSES

A weight-of-evidence approach such as that provided in EPA's RfC Methodology (U.S. EPA, 1994) or in EPA's proposed guidelines for carcinogen risk assessment (U.S. EPA, 1999a) should be used in assessing the database for an agent. This approach requires a critical evaluation of the entire body of available data for consistency and biological plausibility. Potentially relevant studies should be judged for quality and studies of high quality given much more weight than those of lower quality. When both epidemiological and experimental data are available, similarity of effects between humans and animals is given more weight. If the mechanism or mode of action is well characterized, this information is used in the interpretation of observed effects in either human or animal studies. Weight of evidence is not to be interpreted as simply tallying the number of positive and negative studies, nor does it imply an averaging of the doses or exposures identified in individual studies that may be suitable as points of departure (PODs) for risk assessment. The study or studies used for the POD are identified by an informed and expert evaluation of all the available evidence (EPA 2002b, Pp 4-11 to 4-12).

## GUIDELINES FOR CARCINOGEN RISK ASSESSMENT

The cancer guidelines emphasize the importance of weighing all of the evidence in reaching conclusions about the human carcinogenic potential of agents. This is accomplished in a single integrative step after assessing all of the individual lines of evidence, which is in contrast to the step-wise approach in the 1986 cancer guidelines. Evidence considered includes tumor findings, or lack thereof, in humans and laboratory animals; an agent's chemical and physical properties; its structure-activity relationships (SARs) as compared with other carcinogenic agents; and studies addressing potential carcinogenic processes and mode(s) of action, either *in vivo* or *in vitro*. Data from epidemiologic studies are generally preferred for characterizing human cancer hazard and risk. However, all of the information discussed above could provide valuable insights into the possible mode(s) of action and likelihood of human cancer hazard and risk. The cancer guidelines recognize the growing sophistication of research methods, particularly in their ability to reveal the modes of action of carcinogenic agents at cellular and subcellular levels as well as toxicokinetic processes.

Weighing of the evidence includes addressing not only the likelihood of human carcinogenic effects of the agent but also the conditions under which such effects may be expressed, to the extent that these are revealed in the toxicological and other biologically important features of the agent.

The weight of evidence narrative to characterize hazard summarizes the results of the hazard assessment and provides a conclusion with regard to human carcinogenic potential. The narrative explains the kinds of evidence available and how they fit together in drawing conclusions, and it points out significant issues/strengths/limitations of the data and conclusions. Because the narrative also summarizes the mode of action information, it sets the stage for the discussion of the rationale underlying a recommended approach to dose-response assessment.

In order to provide some measure of clarity and consistency in an otherwise free-form, narrative characterization, standard descriptors are used as part of the hazard narrative to express the conclusion regarding the weight of evidence for carcinogenic hazard potential. There are five recommended standard hazard descriptors: *"Carcinogenic to Humans," "Likely to Be Carcinogenic to Humans," "Suggestive Evidence of Carcinogenic Potential," "Inadequate Information to Assess Carcinogenic Potential,"* and *"Not Likely to Be Carcinogenic to Humans."* Each standard descriptor may be applicable to a wide variety of data sets and weights of evidence and is presented only in the context of a weight of evidence narrative. Furthermore, as described in Section 2.5 of these cancer guidelines, more than one conclusion may be reached for an agent (EPA 2005b, Pp 1-11 to 1-12).

The *weight of evidence narrative* is a short summary (one to two pages) that explains an agent's human carcinogenic potential and the conditions that characterize its expression. It should be sufficiently complete to be able to stand alone, highlighting the key issues and decisions that were the basis for the evaluation of the agent's potential hazard. It should be sufficiently clear and transparent to be useful to risk managers and non-expert readers. It may be useful to summarize all of the significant components and conclusions in the first paragraph of the narrative and to explain complex issues in more depth in the rest of the narrative.

The weight of the evidence should be presented as a narrative laying out the complexity of information that is essential to understanding the hazard and its dependence on the quality, quantity, and type(s) of data available, as well as the circumstances of exposure or the traits of an exposed population that may be required for expression of cancer. For example, the narrative can clearly state to what extent the determination was based on data from human exposure, from animal experiments, from some combination of the two, or from other data. Similarly, information on mode of action can specify to what extent the data are from *in vivo* or *in vitro* exposures or based on similarities to other chemicals. The extent to which an agent's mode of action occurs only on reaching a minimum dose or a minimum duration should also be presented. A hazard might also be expressed disproportionately in individuals possessing a specific gene; such characterizations may follow from a better understanding of the human genome. Furthermore, route of exposure should be used to qualify a hazard if, for example, an agent is not absorbed by some routes. Similarly, a hazard can be attributable to exposures during a susceptible lifestage on the basis of our understanding of human development.

The weight of evidence-of-evidence narrative should highlight:

- the quality and quantity of the data;
- all key decisions and the basis for these major decisions; and
- any data, analyses, or assumptions that are unusual for or new to EPA.

To capture this complexity, a weight of evidence narrative generally includes

- conclusions about human carcinogenic potential (choice of descriptor(s), described below),
- a summary of the key evidence supporting these conclusions (for each descriptor used), including information on the type(s) of data (human and/or animal, *in vivo* and/or *in vitro*) used to support the conclusion(s),
- available information on the epidemiologic or experimental conditions that characterize expression of carcinogenicity (e.g., if carcinogenicity is possible only by one exposure route or only above a certain human exposure level),
- a summary of potential modes of action and how they reinforce the conclusions,
- indications of any susceptible populations or lifestages, when available, and
- a summary of the key default options invoked when the available information is inconclusive.

To provide some measure of clarity and consistency in an otherwise free-form narrative, the weight of evidence descriptors are included in the first sentence of the narrative. Choosing a descriptor is a matter of judgment and cannot be reduced to a formula. Each descriptor may be applicable to a wide variety of potential data sets and weights of evidence. These descriptors and narratives are intended to permit sufficient flexibility to accommodate new scientific understanding and new testing methods as they are developed and accepted by the scientific community and the public. Descriptors represent points along a continuum of evidence; consequently, there are gradations and borderline cases that are clarified by the full narrative. Descriptors, as well as an introductory paragraph, are a short summary of the complete narrative that preserves the complexity that is an essential part of the hazard characterization. Users of these cancer guidelines and of the risk assessments that result from the use of these cancer guidelines should consider the entire range of information included in the narrative rather than focusing simply on the descriptor.

In borderline cases, the narrative explains the case for choosing one descriptor and discusses the arguments for considering but not choosing another.

For example, between "suggestive" and "likely" or between "suggestive" and "inadequate," the explanation clearly communicates the information needed to consider appropriately the agent's carcinogenic potential in subsequent decisions.

Multiple descriptors can be used for a single agent, for example, when carcinogenesis is dose- or route-dependent. For example, if an agent causes point-of-contact tumors by one exposure route but adequate testing is negative by another route, then the agent could be described as likely to be carcinogenic by the first route but not likely to be carcinogenic by the second. Another example is when the mode of action is sufficiently understood to conclude that a key event in tumor development would not occur below a certain dose range. In this case, the agent could be described as likely to be carcinogenic above a certain dose range but not likely to be carcinogenic below that range.

Descriptors can be selected for an agent that has not been tested in a cancer bioassay if sufficient other information, e.g., toxicokinetic and mode of action information, is available to make a strong, convincing, and logical case through scientific inference. For example, if an agent is one of a well-defined class of agents that are understood to operate through a common mode of action and if that agent has the same mode of action, then in the narrative the untested agent would have the same descriptor as the class. Another example is when an untested agent's effects are understood to be caused by a human metabolite, in which case in the narrative the untested agent could have the same descriptor as the metabolite. As new testing methods are developed and used, assessments may increasingly be based on inferences from toxicokinetic and mode of action information in the absence of tumor studies in animals or humans.

When a well-studied agent produces tumors only at a point of initial contact, the descriptor generally applies only to the exposure route producing tumors unless the mode of action is relevant to other routes. The rationale for this conclusion would be explained in the narrative.

When tumors occur at a site other than the point of initial contact, the descriptor generally applies to all exposure routes that have not been adequately tested at sufficient doses. An exception occurs when there is convincing information, e.g., toxicokinetic data that absorption does not occur by another route.

When the response differs qualitatively as well as quantitatively with dose, this information should be part of the characterization of the hazard. In some cases reaching a certain dose range can be a precondition for effects to occur, as when cancer is secondary to another toxic effect that appears only above a certain dose. In other cases exposure duration can be a precondition for hazard if effects occur only after exposure is sustained for a certain duration. These considerations differ from the issues of relative absorption or potency at different dose levels because they may represent a discontinuity in a dose-response function.

When multiple bioassays are inconclusive, mode of action data are likely to hold the key to resolution of the more appropriate descriptor. When bioassays are few, further bioassays to replicate a study's results or to investigate the potential for effects in another sex, strain, or species may be useful.

When there are few pertinent data, the descriptor makes a statement about the database, for example, "Inadequate Information to Assess Carcinogenic Potential," or a database that provides "Suggestive Evidence of Carcinogenic Potential." With more information, the descriptor expresses a conclusion about the agent's carcinogenic potential to humans. If the conclusion is positive, the agent could be described as "Likely to Be Carcinogenic to Humans" or, with strong evidence, "Carcinogenic to Humans." If the conclusion is negative, the agent could be described as "Not Likely to Be Carcinogenic to Humans."

Although the term "likely" can have a probabilistic connotation in other contexts, its use as a weight of evidence descriptor does not correspond to a quantifiable probability of whether the chemical is carcinogenic. This is because the data that support cancer assessments generally are not suitable for numerical calculations of the probability that an agent is a carcinogen. Other health agencies have expressed a comparable weight of evidence using terms such as "Reasonably Anticipated to Be a Human Carcinogen" (NTP) or "Probably Carcinogenic to Humans" (International Agency for Research on Cancer).

The following descriptors can be used as an introduction to the weight of evidence narrative. The examples presented in the discussion of the descriptors are illustrative. The examples are neither a checklist nor a limitation for the descriptor. The complete weight of evidence narrative, rather than the descriptor alone, provides the conclusions and the basis for them.

*"Carcinogenic to Humans"*

This descriptor indicates strong evidence of human carcinogenicity. It covers different combinations of evidence.

- This descriptor is appropriate when there is convincing epidemiologic evidence of a causal association between human exposure and cancer.

- Exceptionally, this descriptor may be equally appropriate with a lesser weight of epidemiologic evidence that is strengthened by other lines of evidence. It can be used when all of the following conditions are met: (a) there is strong evidence of an association between human exposure and either cancer or the key precursor events of the agent's mode of action but not enough for a causal association, and (b) there is extensive evidence of carcinogenicity in animals, and (c) the mode(s) of carcinogenic action and associated key precursor events have been identified in animals, and (d) there is strong evidence that the key precursor events that precede the cancer response in animals are anticipated to occur in humans and progress to tumors, based on available biological information. In this case, the narrative includes a summary of both the experimental and epidemiologic information on mode of action and also an indication of the relative weight that each source of information carries, e.g., based on human information, based on limited human and extensive animal experiments.

### *"Likely to Be Carcinogenic to Humans"*

This descriptor is appropriate when the weight of the evidence is adequate to demonstrate carcinogenic potential to humans but does not reach the weight of evidence for the descriptor "Carcinogenic to Humans." Adequate evidence consistent with this descriptor covers a broad spectrum. As stated previously, the use of the term "likely" as a weight of evidence descriptor does not correspond to a quantifiable probability. The examples below are meant to represent the broad range of data combinations that are covered by this descriptor; they are illustrative and provide neither a checklist nor a limitation for the data that might support use of this descriptor. Moreover, additional information, e.g., on mode of action, might change the choice of descriptor for the illustrated examples. Supporting data for this descriptor may include:

- an agent demonstrating a plausible (but not definitively causal) association between human exposure and cancer, in most cases with some supporting biological, experimental evidence, though not necessarily carcinogenicity data from animal experiments;
- an agent that has tested positive in animal experiments in more than one species, sex, strain, site, or exposure route, with or without evidence of carcinogenicity in humans;
- a positive tumor study that raises additional biological concerns beyond that of a statistically significant result, for example, a high degree of malignancy, or an early age at onset;
- a rare animal tumor response in a single experiment that is assumed to be relevant to humans; or
- a positive tumor study that is strengthened by other lines of evidence, for example, either plausible (but not definitively causal) association between human exposure and cancer or evidence that the agent or an important metabolite causes events generally known to be associated with tumor formation (such as DNA reactivity or effects on cell growth control) likely to be related to the tumor response in this case.

### *"Suggestive Evidence of Carcinogenic Potential"*

This descriptor of the database is appropriate when the weight of evidence is suggestive of carcinogenicity; a concern for potential carcinogenic effects in humans is raised, but the data are judged not sufficient for a stronger conclusion. This descriptor covers a spectrum of evidence associated with varying levels of concern for carcinogenicity, ranging from a positive cancer result in the only study on an agent to a single positive cancer result in an extensive database that includes negative studies in other species. Depending on the extent of the database, additional studies may or may not provide further insights. Some examples include:

- a small, and possibly not statistically significant, increase in tumor incidence observed in a single animal or human study that does not reach the weight of evidence for the descriptor "Likely to Be Carcinogenic to Humans." The study generally would not be contradicted by other studies of equal quality in the same population group or experimental system (see discussions of *conflicting evidence* and *differing results*, below);
- a small increase in a tumor with a high background rate in that sex and strain, when there is some but insufficient evidence that the observed tumors may be due to intrinsic factors that cause background tumors and not due to the agent being assessed. (When there is a high background rate of a specific tumor in animals of a particular sex and strain, then there may be biological factors operating independently of the agent being assessed that could be responsible for the development of the observed tumors.) In this case, the reasons for determining that the tumors are not due to the agent are explained;
- evidence of a positive response in a study whose power, design, or conduct limits the ability to draw a confident conclusion (but does not make the study fatally flawed), but where the carcinogenic potential is strengthened by other lines of evidence (such as structure-activity relationships); or
- a statistically significant increase at one dose only, but no significant response at the other doses and no overall trend.

### *"Inadequate Information to Assess Carcinogenic Potential"*

This descriptor of the database is appropriate when available data are judged inadequate for applying one of the other descriptors. Additional studies generally would be expected to provide further insights. Some examples include:

- little or no pertinent information;
- conflicting evidence, that is, some studies provide evidence of carcinogenicity but other studies of equal quality in the same sex and strain are negative. *Differing results*, that is, positive results in some studies and negative results in one or more different experimental systems, do not constitute *conflicting evidence*, as the term is used here. Depending on the overall weight of evidence, differing results can be considered either suggestive evidence or likely evidence; or negative results that are not sufficiently robust for the descriptor, "Not Likely to Be Carcinogenic to Humans."

Appendix B

- negative results that are not sufficiently robust for the descriptor, "Not Likely to Be Carcinogenic to Humans."

*"Not Likely to Be Carcinogenic to Humans"*

This descriptor is appropriate when the available data are considered robust for deciding that there is no basis for human hazard concern. In some instances, there can be positive results in experimental animals when there is strong, consistent evidence that each mode of action in experimental animals does not operate in humans. In other cases, there can be convincing evidence in both humans and animals that the agent is not carcinogenic. The judgment may be based on data such as:

- animal evidence that demonstrates lack of carcinogenic effect in both sexes in well designed and well-conducted studies in at least two appropriate animal species (in the absence of other animal or human data suggesting a potential for cancer effects),
- convincing and extensive experimental evidence showing that the only carcinogenic effects observed in animals are not relevant to humans,
- convincing evidence that carcinogenic effects are not likely by a particular exposure route (see Section 2.3), or
- convincing evidence that carcinogenic effects are not likely below a defined dose range. A descriptor of "not likely" applies only to the circumstances supported by the data. For example, an agent may be "Not Likely to Be Carcinogenic" by one route but not necessarily by another. In those cases that have positive animal experiment(s) but the results are judged to be not relevant to humans, the narrative discusses why the results are not relevant.

*Multiple Descriptors*

More than one descriptor can be used when an agent's effects differ by dose or exposure route. For example, an agent may be "Carcinogenic to Humans" by one exposure route but "Not Likely to Be Carcinogenic" by a route by which it is not absorbed. Also, an agent could be "Likely to Be Carcinogenic" above a specified dose but "Not Likely to Be Carcinogenic" below that dose because a key event in tumor formation does not occur below that dose (EPA 2005b, Pp 2-49 to 2-58).

## A FRAMEWORK FOR ASSESSING HEALTH RISKS OF ENVIRONMENTAL EXPOSURES TO CHILDREN

The WOE approach requires a critical evaluation (expert judgment) of all available data for consistency and biological plausibility. Criteria for this assessment are not presented here; rather, considerations important for the WOE are described. The key to WOE conclusions is the provision of a clear justification for decisions. Finally, the extent of the database is summarized, and assumptions made in the assessment are explicitly detailed. Further details about EPA's WOE approach can be found in the *Methods for Derivation of Inhalation Reference Concentrations and Application of Inhalation Dosimetry* (U.S. EPA, 1994), *Guidelines for Carcinogen Risk Assessment* (U.S. EPA, 2005b), and *Supplemental Guidance for Assessing Cancer Susceptibility from Early Life Exposure to Carcinogens* (U.S. EPA, 2005c). *A Review of the Reference Dose and Reference Concentration Processes* (U.S. EPA, 2002b, Section 4.3.2.1.) and *Determination of the Appropriate FQPA Safety Factor(s) on Tolerance Assessment* (U.S. EPA, 2002c, Section III) provide additional detail on the WOE.

Key themes for the consideration of toxicity data in a WOE assessment, as adapted from Gray et al. (2001), are shown in Figure 4-5. This figure focuses on judging animal studies within a WOE assessment. However, if adequate human studies are available they would be given more weight. The process for evaluating these considerations is described in the following subsections. In this process, the quality of potentially relevant studies is judged, modifiers and interactions are detailed, outcomes across species are compared, TK and TD data are examined and weighed for comparisons across species, and the uncertainties and data gaps are determined. SARs with other chemicals or chemical classes are explored to determine the extent to which these data can inform the assessment via an MOA discussion or reduce uncertainties.

## GUIDELINES FOR NEUROTOXICITY RISK ASSESSMENT

The interpretation of data as indicative of a potential neurotoxic effect involves the evaluation of the validity of the database. This approach and these terms have been adapted from the literature on human psychological testing (Sette, 1987; Sette and MacPhail, 1992), where they have long been used to evaluate the level of confidence in different measures of intelligence or other abilities, aptitudes, or feelings. There are four principal questions that should be addressed: whether the effects result from exposure (content validity); whether the effects are adverse or toxicologically significant (construct validity); whether there are correlative measures among behavioral, physiological, neurochemical, and morphological endpoints (concurrent validity); and whether the effects are predictive of what will happen under various conditions (predictive validity). Addressing these issues can provide a useful framework for evaluating either human or animal studies or the weight of evidence for a chemical (Sette, 1987; Sette and MacPhail, 1992). The next sections indicate the extent to which chemically induced changes can be interpreted as providing evidence of neurotoxicity.

The qualitative characterization of neurotoxic hazard can be based on either human or animal data (Anger, 1984; Reiter, 1987; U.S. EPA, 1994). Such data can result from accidental, inappropriate, or controlled experimental exposures. This section describes many of the general and some of the specific characteristics of human studies and reports of neurotoxicity. It then describes some features of animal studies of neuroanatomical, neurochemical, neurophysiological, and behavioral effects relevant to risk assessment. The process of characterizing the sufficiency or insufficiency of neurotoxic effects for risk assessment is described in section 3.3. Additional sources of information relevant to hazard characterization, such as comparisons of molecular structure among compounds and in vitro screening methods, are also discussed.

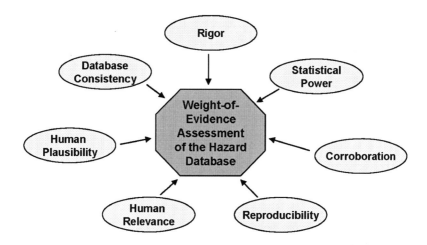

**FIGURE 4-5** Conceptual view of a weight of evidence (WOE) assessment. This figure illustrates the critical considerations within a WOE assessment of toxicity data. *Rigor* is the degree of proper conduct and analysis of a study; greater weight is generally given to more rigorous studies. *Statistical Power* is the ability of a study to detect effects of a given magnitude. *Corroboration* means that specific effects are replicated in similar studies, similar effects are observed under varied conditions and /or similar effects are observed in multiple laboratories. *Reproducibility* means that an effect is observed in multiple species by various routes of exposure. *Relevance to Humans* means that similar effects are observed in humans or in a species taxonomically related to humans or at doses similar to those expected in humans. *Plausibility to Humans* is the determination of whether a similar metabolism, mechanisms of damage and repair, and molecular target of response could be expected to occur in humans, based on an evaluation of the biologic mechanism of a toxic response in animals. *Database Consistency* is the extent to which all of the data are similar in outcome and dose (exposure-response) and are operating under a single biologically plausible assumption (mode of action). Source: Adapted from Gray et al. 2001, EPA 2006, Pp 29-30.

The hazard characterization should:

a. Identify strengths and limitations of the database:
   - Epidemiological studies (case reports, cross-sectional, case-control, cohort, or human laboratory exposure studies);
   - Animal studies (including structural or neuropathological, neurochemical, neurophysiological, behavioral or neurological, or developmental endpoints).

b. Evaluate the validity of the database:
   - Content validity (effects result from exposure);
   - Construct validity (effects are adverse or toxicologically significant);
   - Concurrent validity (correlative measures among behavioral, physiological, neurochemical, or morphological endpoints);
   - Predictive validity (effects are predictive of what will happen under various conditions).

c. Identify and describe key toxicological studies.

d. Describe the type of effects:
   - Structural (neuroanatomical alternations);
   - Functional (neurochemical, neurophysiological, behavioral alterations).

e. Describe the nature of the effects (irreversible, reversible, transient, progressive, delayed, residual, or latent).

f. Describe how much is known about how (through what biological mechanism) the chemical produces adverse effects.

g. Discuss other health endpoints of concern.

h. Comment on any nonpositive data in humans or animals.

I. Discuss the dose-response data (epidemiological or animal) available for further dose-response analysis.

j. Discuss the route, level, timing, and duration of exposure in studies demonstrating neurotoxicity as compared to expected human exposures.

k. Summarize the hazard characterization:
   - Confidence in conclusions;
   - Alternative conclusions also supported by the data;
   - Significant data gaps; and
   - Highlights of major assumptions.

## REFERENCES

American Thoracic Society. 1985. Guidelines as to what constitutes an adverse respiratory health effect, with special reference to epidemiologic studies of air pollution. Am. Rev. Respir. Dis. 131(4):666-668.

Anger, W.K. 1984. Neurobehavioral testing of chemicals: Impact on recommended standards. Neurobehav. Toxicol. Teratol. 6(2):147-153.

EPA (U.S. Environmental Protection Agency). 1986. Guidelines for Mutagenicity Risk Assessment. U.S. Environmental Protection Agency [online]. Available: http://www.epa.gov/osa/mmoaframework/pdfs/MUTAGEN2.PDF [accessed Nov. 19, 2010].

EPA (U.S. Environmental Protection Agency). 1991. Guidelines for Developmental Toxicity Risk Assessment. U.S. Environmental Protection Agency [online]. Available: http://cfpub.epa.gov/ncea/cfm/recordisplay.cfm?deid=23162 [accessed Nov. 19, 2010].

EPA (U.S. Environmental Protection Agency). 1994. Methods for Derivation of Inhalation Reference Concentrations and Application of Inhalation Dosimetry. U.S. Environmental Protection Agency [online]. Available: http://www.epa.gov/raf/publications/methods-derivation-inhalation-ref.htm [accessed Nov. 19, 2010].

EPA (U.S. Environmental Protection Agency). 1998. Guidelines for Neurotoxicity Risk Assessment. U.S. Environmental Protection Agency [online]. Available: http://www.epa.gov/raf/publications/pdfs/NEUROTOX.PDF [accessed Dec. 16, 2010].

EPA (U.S. Environmental Protection Agency). 1999a. Guidelines for Carcinogen Risk Assessment [Review Draft]. NCEA-F-0644. Risk Assessment Forum, U.S. Environmental Protection Agency, Washington, DC. July 1999 [online]. Available: http://cfpub.epa.gov/ncea/cfm/recordisplay.cfm?deid=54932#Download [accessed Mar. 17, 2011].

EPA (U.S. Environmental Protection Agency). 2002b. A Review of the Reference Dose and Reference Concentration Process. U.S. Environmental Protection Agency [online]. Available: http://www.epa.gov/iris/RFD_FINAL1.pdf [accessed Nov. 19, 2010].

EPA (U.S. Environmental Protection Agency). 2002c. Determination of the Appropriate FQPA Safety Factor(s) in Tolerance Assessment. Office of Pesticide Programs, U.S. Environmental Protection Agency, Washington, DC. February 28, 2002 [online]. Available: http://www.epa.gov/oppfead1/trac/science/determ.pdf [accessed Mar. 17, 2011].

EPA (U.S. Environmental Protection Agency). 2005b. Guidelines for Carcinogen Risk Assessment. U.S. Environmental Protection Agency [online]. Available: http://www.epa.gov/osa/mmoaframework/pdfs/CANCER-GUIDELINES-FINAL-3-25-05%5B1%5D.pdf [accessed Nov. 19, 2010].

EPA (U.S. Environmental Protection Agency). 2005c. Supplemental Guidance for Assessing Cancer Susceptibility from Early-Life Exposure to Carcinogens. EPA/630/R-03/003F. Risk Assessment Forum, U.S. Environmental Protection Agency, Washington, DC. March 2005 [online]. Available: http://www.epa.gov/ttn/atw/childrens_supplement_final.pdf [accessed Mar. 17, 2011].

EPA (U.S. Environmental Protection Agency). 2006. A Framework for Assessing Health Risks of Environmental Exposures to Children. U.S. Environmental Protection Agency [online]. Available: http://cfpub.epa.gov/ncea/cfm/recordisplay.cfm?deid=158363 [accessed Nov. 19, 2010].

Gray, G.M., S.I. Baskin, G. Charnley, J.T. Cohen, L.S. Gold, N.I. Kerkvliet, H.M. Koenig, S.C. Lewis, R.M. McClain, L.R. Rhomberg, J.W. Snyder, and L.B. Weekley. 2001. The Annapolis accords on the use of toxicology in risk assessment and decision-making: An Annapolis Center Workshop report. Toxicol. Mech. Methods 11(3):225-231.

Perlin, S.A., and C. McCormack. 1988. Using weight-of-evidence classification schemes in the assessment of non-cancer health risks. Pp. 482-486 in Proceedings of the 5th National Conference on Hazardous Wastes and Hazardous Materials (HWHM '88), April 19-21, Las Vegas, NV. Springfield, MD: Hazardous Materials Control Research Institute.

Reiter, L.W. 1987. Neurotoxicology in regulation and risk assessment. Dev. Pharmacol. Ther. 10(5):354-368.

Russell, L.B., C.S. Aaron, F. de Serres, W.M. Generoso, K.L. Kannan, M. Shelby, J. Springer, and P. Voytek. 1984. A report of the U.S. Environmental Protection Agency Gene-Tox Program. Evaluation of mutagenicity assays for purposes of genetic risk assessment. Mutat. Res. 134(2-3):143-157.

Sette, W.F. 1987. Complexity of neurotoxicological assessment. Neurotoxicol. Teratol 9(6):411-416.

Sette, W.F., and R.C. MacPhail. 1992. Qualitative and quantitative issues in assessment of neurotoxic effects. Pp. 345-361 in Neurotoxicology, 2nd Ed, H.A. Tilson, and C. Mitchell, eds. Target Organ Toxicity Series. New York: Raven Press.

SOT (Society of Toxicology). 1982. Animal data in hazard evaluation: Paths and pitfalls. Task Force of Past Presidents. Fundam. Appl. Toxicol. 2(3):101-107.